Indian Ocean Resources and Technology

Indian Ocean Resources and Technology

G. S. Roonwal

CRC Press
Taylor & Francis Group
Boca Raton London New York

CRC Press is an imprint of the
Taylor & Francis Group, an **informa** business

CRC Press
Taylor & Francis Group
6000 Broken Sound Parkway NW, Suite 300
Boca Raton, FL 33487-2742

First issued in paperback 2020

ISBN-13: 978-0-367-57277-8 (pbk)
ISBN-13: 978-1-138-09534-2 (hbk)

Library of Congress Cataloging-in-Publication Data

Names: Roonwal, G. S. (Ganpat Singh), 1940- author.
Title: Indian Ocean resources and technology / Ganpat Singh Roonwal.
Description: Boca Raton : Taylor & Francis, CRC Press, 2018. | Includes bibliographical references and index.
Identifiers: LCCN 2017023686| ISBN 9781138095342 (hardback : alk. paper) | ISBN 9781315105697 (ebook)
Subjects: LCSH: Marine mineral resources--Indian Ocean.
Classification: LCC TN264 .R665 2018 | DDC 333.8/509165--dc23
LC record available at https://lccn.loc.gov/2017023686

Visit the Taylor & Francis Web site at
http://www.taylorandfrancis.com

and the CRC Press Web site at
http://www.crcpress.com

To my parents and family who gave me this wonderful world, affectionate

upbringing, discipline, hard work, and sense of humor

To my teachers who gave me a desire to learn

Contents

List of Figures

List of Tables

Preface

In 1981 Daniel Behrman, summarizing the achievements of the International Ocean Expedition (1959–65), called the Indian Ocean "the largest unknown." From this, after six decades, the Indian Ocean today has become an ocean of competition, especially the seabed resources (Roonwal 2015). It is, therefore, timely to have an assessment of the resources of the Indian Ocean and the technology needed for good utilization of these for the benefit of the people around the Indian Ocean. Clearly, India could play a major role in initiating such an activity. To both utilize a 73,426,500 sq km of water mass, one-seventh of the surface of the earth. The population of more than 2 billion would be benefited from this approach. That is the spirit of UN Conference on the Law of the Sea (UNCLOS) that the oceans could be the "common heritage of mankind." This would be a tribute to the visionaries like Paul Tchernia, Lloyd Berkner, Roger Revelle, George Deacon, Robert G. Snider, Gunter Dietrich, Maurice Ewing, Anton Brunn, and many others for initiating the International Indian Ocean Expedition (IIOE).

In view of India's rapid progress in all area of science and technology, and in particular in utilizing and managing its marine resources under the new ocean regime, it is timely that an assessment is done. It is the aim of this book to provide an overview of the area and to highlight the potential market opportunities represented by this vast and rapidly developing nation. In doing so the following aspects have been covered.

- Mineral resources (hard minerals oil and gas), deep sea. The shallow sea and the coastal zone minerals
- Living resources (fish and aquaculture)
- Energy resources (OTEC, wave, tidal) and freshwater from the sea
- Technology and economics of deep-sea mining
- Environment, pollution, and waste disposal

India has long viewed its path of development as indicated by advances in science and technology. Therefore, its market potential has aroused interest in many industrialized advanced nations. As a result of open market policy of the government, opportunity has opened for great potential and possibility. In doing so, India has undergone a rapid expansion of marine science and technology, which resulted in the establishment of research institutions such as the NIO, NCAOR, NIOT, and an expansion of marine science activity by GSI, ONGC, and others. Several research ships are available both for deep sea and shallow survey and research as we witness science grow more diverse and an inevitable increase in the degree to which specialization isolates the practitioners to often ever large number of subfields. This may lead to barriers, often impenetrable. Therefore, an attempt to bring an overview assessment of both the deep sea the shallow sea, the EEZ, and the coastal zone all in a single place shall be of immense utility to all—the active researchers, policymakers, the technology and engineering consortia, and students. In fact, this book will not only fill the gap and provide useful reference work but it will also generate future discussion concerning the most appropriate strategy for further development in the technology and role of India in utilizing the Indian Ocean for the benefit of the people.

From the time when sustainable development goals (SDG) are being recognized, more and more attention is focused on an inclusive use of ocean—the body itself and the resources both living and nonliving the ocean offers. In addition, ocean currents, salinity variation, temperature variants, and even the tidal waves are being considered for utilization as a source of renewable energy. I realized early the significance of the Indian Ocean, an ocean that was not so well studied as the Pacific and the Atlantic Oceans. The International Indian Ocean Expedition (1959–65) gave a big push to the science of ocean in this region. It led to creation of ocean research institutes in the region. In India, the National Institute of Oceanography (NIO) was established, an extension and result of the IIOE, to continue work on huge observation data, and samples collected during the IIOE. Mero's (1965) book *Mineral Resources of the Sea* created added interest in the seafloor minerals. Therefore, it was appropriate for me to write *The Indian Ocean Exploitable Mineral and Petroleum Resources* (1986). The access to resources in the Indian Ocean has created a delicate situation. There is now great interest in seabed minerals. Here in addition to India, others who have signed contracts for exploration include South Korea, China, and Germany. It is resulting in a sort of competition (Roonwal, 2015). The prospect of ocean economy for a holistic development of the Indian Ocean and all its resource potential needs to be understood. Hopefully a good synthesis and scientific data interpretation shall prevent the Indian Ocean from degradation.

This book is an attempt in this direction. There may be several other aspects that could have been included, but we have to wait for more data and synthesis. After all it's the beginning of the journey, a long journey of humanity and welfare. It is in a way the "first step" in this journey, and I look forward to support and suggestions in this attempt.

Acknowledgments

My sincere thanks to Dr. K. Kanjilal, director, and Dr. Sundeep Chopra, head of AMS Group, Inter-University Accelerator Centre (IUAC), New Delhi, for faculties and support. I also thank Narender Kumar who brought my text in the computer format and Amerjeet Roop Rai for drafting of the line diagrams and figures. I thank Dr. Pankaj Kumar, Rajeev Sharma, and Pryamboda Nayak, librarian of IUAC, for various kinds of help.

I am grateful to Prashant, Geetu, Kabir, and Kanav Behki's for family support to complete this work.

Manju Bala Narender Kumar was always a support as we spent long hours at work on computers.

My sincere thanks to Professor David J. Sanderson, University of Southampton, UK, and Professor John Wiltshire of University of Hawaii for their suggestions in the final evolution of the theme. I thank Dr. G. A. Ramadass, Dr. K. Gopakumar, and Dr. Purnima Jalihal of the National Institute of Ocean Technology, Chennai, for discussions on offshore technology regarding freshwater and energy from the sea; Dr. Sanjay Kali of IMMT Bhubaneswar for the photograph of cobalt-rich crust; and Prashant Dubey of ONGC for discussions.

I also received the benefit of discussions with several distinguished participants during the eight Deep-sea Mining Summits in Singapore and London in 2015–16, especially Dr. Frank Lim (2H). Paul Holthus of World Ocean Council, Honolulu, United States, Engineer Henk van Muijen of IHC Mining, the Netherlands, Ray Wood, Chatham Rock Phosphate, New Zealand, and Phil Lotto Marine Space.

I was inspired and introduced to marine minerals by Professor G. H. W. Friedrich (since deceased) of the Institute of Mineralogy and Economic Geology RWTH Aachen, Germany. I had the privilege of sailing with him for manganese nodule exploration. I sailed together with Dr. Vesna Machig, Hanover, Germany, in the exploration campaign for seafloor sulfide and several visits to Hanover and Delhi for collaboration. Professor P. A. Rona (since deceased), and Dr. J. D. Hein of the USGS engaged me in a discussion on cobalt-rich crust. I am grateful to Professor S. D. Scott, Toronto, Canada, for discussion, support, and sailing together for exploration of seafloor sulfide; Professor P. Halbach, Berlin, for discussion on seabed minerals in Germany and Delhi; Professor D. S. Cronan, Imperial College, London, for several discussions in London and Delhi and sailing together; Dr. G. P. Glasby, Sheffield and John Yates, Manchester, for exciting discussions on marine minerals during their visits to Delhi; and Dr. J. S. Chung of ISOPE, United States, for useful discussions during ISOPE meetings.

I thank the different publishing houses and authors for supply of information and permission to reproduce or modify requested figures for this book. Centre for Marine Living Resources and Ecology (MoES), and Dr. M. P. Wakdikar of MoES for photographs of fishes of the Indian Ocean; Dr. Rajiv Nigam of the National Institute of Oceanography of Goa for discussions and photographs of fishes of the Indian Ocean; D. Singh, CMD, IRE Ltd.; A. K. Mohapatra, and Dr. B. Mishra, Indian Rare Earths (OSCOM, Chhatrapur) for samples of mining operations for heavy mineral sand on Gopalpur coast; and Oil and Natural Gas Corporation (ONGC), India, for the photograph of offshore platforms.

The preparation of this book has been facilitated by a grant given by the Hindustan Zinc Ltd. (a unit of Vedanta Group). My sincere thanks are to the chief executive officer, Sunil

Duggal, and the chief operating officer (Mines), L. S. Sekhawat for understanding and support.

The editorial group at the New Delhi office of Taylor & Francis Group/CRC Press, has been helpful at all stages. Special thanks to Dr. Gagandeep Singh and Mouli Sharma for their support.

Author

G. S. Roonwal is an honorary visiting professor at the Inter-University Accelerator Centre, New Delhi. Earlier, he was a professor of geology and head of the Department of Geology, University of Delhi. He has participated in the exploration of seafloor manganese nodules and the seabed massive sulfides. He has authored *The Indian Ocean Exploitable Mineral and Petroleum Resources* (Springer, Heidelberg) and *Competition for Seabed Resources in the Indian Ocean*. Additionally, he is a coeditor of *India's Exclusive Economic Zone* and *Living Resources of India's Exclusive Economic Zone*. Professor Roonwal is a reputed geologist, a recipient of the National Mineral Award by the Government of India, and several other awards. He has worked with distinguished marine geologists overseas and traveled to several countries adjacent to the oceans. He is listed in Marques *Who's Who in the World* and Marques *Who's Who in Asia* for several years.

Acronyms

2D/3D/4D	two dimensional/three dimensional/four dimensional (time-lapsed)
3D-3C	three dimension–three component
AFERNOD	Association Francais pour L'Etude et Recherche des Nodules
AMR	ArbeitsgemeinschaftmeerestechnischgewinnbareRohstoffe
BCM	billion cubic meters
BGR	Bundesanstalt fur Geowissenschaft und Rohstoffe
BIO	Bedford Institute of Oceanography
BMFT	Bundesministerium-fur Forschung und Technologie
BOPD	barrels of oil per day
BRGM	Bureau de Recherches Geologiques et Minieres
BTOE	billion tonnes oil equivalent
CAAGR	compounded average annual growth rate
CBM	coalbed methane
CLB	continuous line bucket
COGLA	Canada Oil and Gas Lands Administration
CRAM	common reflection angle migration
CSEM	controlled source electro magnetic
CSM	Chantiers sous-Marin
CTD	conductivity, temperature, depth
CZMA	Coastal Zone Management Act
DEIS	draft environmental impact statement
DFN	discreet fraction network
DGH	Directorate General of Hydrocarbons
E&P	exploration and production
EEZ	exclusive economic zone
EOB	Eastern Ocean Basin
EPR	East Pacific Rise
FRM	fluid replacement model
GARIMAS	Galapagos rift massive sulfide
GEMONOD	Groupement D'interet Public pour la Mise au Point des Moyens Necessaires a l'Exploitation des Nodules Polymetalliques
GEOMETEP	Geothermal Metallogenesis East Pacific
GH	gas hydrates
GLORIA	Geological Long-Range Inclined Asdic
GOI	Government of India
GPS	global positioning system
GSI	Geological Survey of India
HELP	Hydrocarbon Exploration Licencing Policy
HOEC	Hindustan Oil Exploration Company

HPB	Heera-Panna-Bassein
HPHT	high pressure–high temperature
IFREMER	Institut Francais de Recherche pour l'Exploitation de la Mer
IOS	Institute of Oceanographic Sciences
JAMSTEC	Japan Marine Science and Technology Centre
KG	Krishna–Godavari
LWD	logging while drilling
MAR	Mid-Atlantic Ridge
MBPD	million barrels per day
MMS	Minerals Management Service
MMSCD/ MMSCMD	million metric standard cubic meters per day
MoES	Ministry of Earth Sciences
MOSES	Magnetometric Offshore Electrical Sounding
MPD	managed pressure drilling
MPL	Marine Physical Laboratory
MPS	Marine Polymetallic Sulphides
MRES	Ministre Delegue Charge de la Recherche et de l'Enseignment Superieur
NANG	nonassociated natural gas
NEC	Northern East Coast
NELP	New Exploration Licensing Policy
NIO	National Institute of Oceanography, Goa, India
NIOT	National Institute of Ocean Technology
NOAA	National Oceanic and Atmospheric Administration
NOC	national oil companies
NSF	National Science Foundation
OCSLA	Outer Continental Shelf Lands Act
OIL	Oil India Ltd.
ONGC	Oil and Natural Gas Corporation Ltd.
OOMA	Deep Ocean Minerals Association
OORD	Deep Ocean Resources Development Co. Ltd.
ORI	Ocean Research Institute
OSU	Oregon State University
PNEHO	Programme National de l'Etude de l'Hydrothermalisme Oceanologique
PST	passive seismic tomography
PSU	public sector undertaking
RIL	Reliance Industries Limited
ROV	remotely operated vehicle
RSA	Reciprocating States Agreement
RSC	Red Sea Commission
RSS	rotary steerable system
SBL	seabed logging

SIO	Scripps Institute of Oceanography
SP	self-potential
TOM	total dissolvable manganese
UNCLS	United Nations Convention on the Law of the Sea
USBM	United States Bureau of Mines
USGS	United States Geological Survey
WHOI	Woods Hole Oceanographic Institute

Conversion Factors

1 meter = 3.281 feet
1 centimeter = 0.39 inches
1 micron = 0.039 thousandths of an inch (1 millionth of a meter)
1 foot = 0.305 meter
1 kilometer = 100 metres = 0.621 miles
1 mile = 1.609 kilometer
1 gram = 0.035 ounce (dry)
1 ounce (dry) = 28.35 gram
1 kilogram = 1000 gram = 2.205 pounds
1 pound = 0.454 kilogram
1 metric ton = 0.984 kilogram long ton (UK) = 1.102 short tons (US)
1 long ton (UK) = 1.016 metric tons = 1.120 short tons (US)
1 short ton (US) = 0.907 metric ton = 0.893 long ton (UK)
1 litre = 0.220 gallon UK = 0.264 gallon (US)
1 gallon UK = 4.546 liter = 1.201 gallon US
1 gallon US = 3.785 liter—0.833 gallon UK
1 barrel (oil) = 0.132 metric tons = 0.134 long tons (UK)
1 barrel (oil) = 0.150 short tons (US) = 159 liters
1 barrel (oil) = 35 gallon (UK) = 42 gallon (US)

High Numerals
1,000,000 = Million
1,000,000,000 = Billion (1000 million in Europe)
1,000,000,000,000 = Trillion (1000 billion in Europe)

Area
10,000 sq meters = 1 hectare (2.471 acres)
1 acre = 0.405 hectare
1 sq mile = 2.599 sq km

Conversion Factors

1 metre = 3.281 feet
1 centimeter = 0.39 inches
1 micron = one thousandth of an inch (1 millionth of a metre)
1 foot = 0.305 metre
1 kilometer = 1000 metres = 0.621 miles
1 mile = 1.609 kilometer
1 gram = 0.035 ounce (Oz)
1 ounce (oz) = 28.35 gram
1 kilogram = 1000 grams = 2.205 pounds
1 pound (lb) = 453 kilogram
1 metric ton = 1000 kilogram long ton (UK) = 1.102 short tons (US)
1 Long ton (UK) = 1.016 metric tons = 1.121 short tons (US)
1 short ton (US) = 0.907 metric ton = 0.893 long ton (UK)
1 litre = 0.220 gallon UK = 0.265 gallon (US)
1 gallon UK = 4.546 litre = 1.201 gallon US
1 gallon US = 3.785 litre = 0.832 gallon UK
1 barrel (oil) = 0.136 metric tons = 0.134 long tons (UK)
1 barrel (oil) = 0.150 short tons (US) = 159 litres
1 barrel (oil) = 35 gallon (UK) = 42 gallon (US)

High Numerals

1,000,000 = Million
1,000,000,000 = Billion (10^9 million in Europe)
1,000,000,000,000 = Trillion (10^9 billion in Europe)

Area

10,000 sq meters = 1 hectare (2.471 acres)
1 acre = 0.405 hectare
1 sq mile = 2.590 sq km

Introduction

This book attempts to bring the different aspects of resources—the minerals, energy, ocean thermal energy conversion (OTEC), tidal, fishes, and others—and combine them with technology needs so that the resources can be utilized for the benefit of society. This book aims to be a one-stop reference source for everyone who is interested in understanding the issues surrounding the Indian Ocean. This book has been carefully updated to include information on key themes and details associated with the Indian Ocean. It is hoped that this book would become a valuable reference for understanding the Indian Ocean. This book has been divided into chapters and sections, making it easy to access the information one would need.

1

The Indian Ocean and Its Associates

1.1 Uniqueness of Indian Ocean

Oceans contain resources like fish, seaweed, and other organisms that provide livelihood to millions of people. Fish and shellfish provide a plentiful supply of proteins to a worldwide population. Seaweed derivatives are used in the production of food, cosmetics, detergents, and industrial lubricants. The chemicals that protect species against predators may serve humanity in combating many of the prevalent diseases such as hypertension, cardiovascular problems, and bacterial and viral infections. In the coming decades, the ocean and its sources will also be an important resource of mineral and energy.

The physical profile from the shoreline to seafloor indicates two major physiographic regions: (a) continental margins, which represent the submerged edges of the continental masses, and (b) the deep ocean basins. These physiographic features have resulted from the geological events that occurred below them (mantle, crustal layers) and huge loads of varying types of sediments deposited over them. Mineral deposits are present in all parts of the ocean. Minerals containing zinc, copper, nickel, and cobalt occur in the mid-oceanic ridge system and on the oceanic plains far away from the continents in addition to deposits of aggregates, phosphorite, and heavy minerals in the shallower to inshore waters.

In the coming decades, we will look increasingly to the oceans for many of our non-renewable mineral resources. No doubt, the pace of this development will fluctuate as mineral prices fall or as deposits of new onshore mineral resources are discovered. At present, oil and gas worth billions of dollars per annum are extracted from the offshore zones. Commercially several mineral commodities from the territorial waters of at least 27 countries of a value of around $600 million per annum are extracted. However, the seafloor is not a host for a vast bonanza of mineral deposits waiting to be readily exploited. The seafloor does not give up its resources easily and the successful exploitation of offshore deposits commonly requires a blend of good science, high technology, adequate capital, a deal of courage, and an element of luck.

Further, India and her associated islands have an extensive influence of oceans. History has shown of flourishing shipbuilding industry trade and commerce, which India carried out with the neighboring countries and other far off countries. Today oceanography is very important to India because the 7500 km long coast line, the real interface of land sea atmosphere, is thickly inhabited. Many large cities of India are located in the coastal zone. The coastal population is expected to double within a few decades. Such large concentration of population and related human activity such as ports will pose several problems because of the threat to the ecological system of the coastal zones. This will demand a proper management and sustainable development of coastal zones in terms of natural disaster, highway construction, impact of port, impact of tourism, disposal of solid waste,

TABLE 1.1

Marine Environment and Associated Minerals

Commodity	Setting	Status	Growth Potential
Aggregates (sand and gravel)	Beach and shallow marine	Operational	High
Cobalt	Deep sea nodules	Nonoperational	Moderate
Cobalt	Crusts on seamounts	Nonoperational	Low
Copper	Deep sea nodules	Nonoperational	Moderate
Copper	Deep sea sulfides	Nonoperational	High (2010)
Diamonds	Shallow marine	Operational	High
Gold	Shallow marine placer	Nonoperational (Some artisanal)	Moderate
Gold	Deep sea sulfides	Nonoperational	High (2010)
Heavy minerals (chromium, rare earths, thorium, titanium, zirconium)	Beach and shallow marine placer	Operational (minor)	Moderate
Lead	Deep sea sulfides	Nonoperational	High
Lime (coral, shells)	Beach	Operational (minor)	Moderate
Methane (gas hydrate)	Shallow/intermediate marine	Nonoperational (test well on land)	Moderate but technological challenges
Nickel	Deep sea nodules	Nonoperational	Moderate
Nickel	Crusts on seamounts	Nonoperational	Low
Phosphate	Shallow marine and seamounts	Nonoperational	Moderate/low
Platinum group metals	Crusts on seamounts	Nonoperational	Low
Rare earth elements	Crusts on seamounts	Nonoperational	Low
Salt	Very shallow marine (evaporation)	Operational	Moderate
Silver	Deep sea sulfides	Nonoperational	High (2010)
Tin	Shallow marine placer	Operational to 50 m water depth	High to depth >80 m water depth
Zinc	Deep sea sulfides	Nonoperational	High (2010)

Source: After Scott, S.D. and committee members, 2008. Mineral deposits in the sea: Second report of the ECOR panel on marine mining, 34pp.

water disposal, aquaculture, and coastal mining and the resultant erosion. Tin and other heavy minerals, lime sand, diamonds, lime, solutes, and precious metals are found in the ocean. Table 1.1 lists the various marine minerals that could be recovered from the sea. Detailed compilations may be found in Roonwal (1986), Lenoble et al., (1995), Cruickshank (1998), Antrim (2005), Hein et al. (2005), and Rona (2008).

India and its associated islands have a very extensive coast line. Subsequent to the declaration of a 200-mile exclusive economic zone (EEZ), India gained sovereignty over two million square kilometers of the sea, an area equal to almost 60% of the total landmass. The EEZ declaration, and its potential use as an ocean management tool, focused attention on marine resources and upon the development of new marine technology. Activity in the oceanographic sciences is increasing within the country, which itself a reflection of the importance attached to the development of marine resources by India.

A country with a strong scientific and academic tradition, India's public sector research and development programs are geared toward economic and industrial success. A workshop

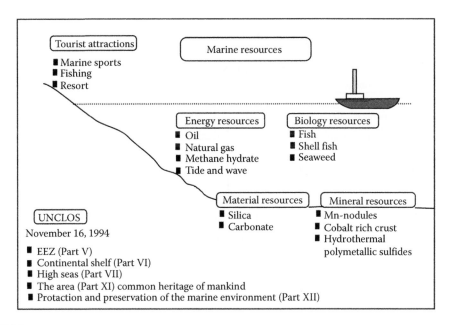

FIGURE 1.1
Oceanic resources.

in 1982 (Inter-govermental Oceanographic Commission, 1982) addressed the problems of the development of ocean sciences in the coming decade (Thiede, 1983). The workshop demonstrated the present sophistication of ocean sciences and pointed out that the development is likely to continue to expand in much the same way as they have since the Second World War. Hsu and Thiede (1992) have given a wide overview of the development and the present status of oceanography. The importance of ocean resource development globally is well illustrated by the negotiations of the Third UN Law of the Sea Conference, which began in 1973, and these have been among the most protracted and complex international negotiations ever held (Amann, 1982; Mann Borgese, 1992) (Figures 1.1 and 1.2).

The aim of this volume is to focus attention on the ocean and seas around India as a region of considerable scientific and economic interest, which is worthy of much more extensive investigation over the coming decades, as India matures from being primarily an agricultural nation to an industrialized one. Contrary to the accepted viewpoint, spending on ocean research in India is not high in comparison to that of advanced industrial nations. As the author has a background in earth sciences, there may be a bias toward this particular area. This is considered inevitable in a multidisciplinary subject, but it does not imply a lesser significance of the other areas of research and development. It is hoped that this volume will stimulate discussion on the future of oceanography and the sustainable management of the ocean surrounding India.

1.2 The Indian Ocean and Subseas

The Indian Ocean is the smallest of the three large oceans, namely the Pacific, the Atlantic, and the Indian Oceans. It covers an area of 28,400,000 sq miles (73,600,000 sq km), and

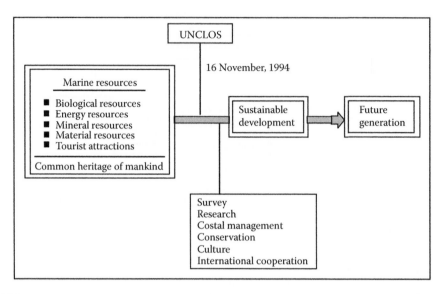

FIGURE 1.2
Law of the sea and resources.

constitutes about one-seventh of the earth's surface (Figure 1.3). The ocean represents about 20% of the total world ocean area and is approximately 9500 km wide between the southern point of Africa and Antarctica. It narrows steadily toward the north, where it is separated by India into the Arabian Sea and Bay of Bengal. Two adjoining seas are connected with the Arabian Sea, the Red Sea (through the Strait of Bab el Mandeb), covering an area of 169,100 sq miles (437,750 sq km) and the Persian Gulf through the Strait of Hormuz (Figure 1.4).

The ocean includes the oceanic region bounded on the north by Iran, Pakistan, India, Bangladesh, and Burma; on the west by the Arabian peninsula and Africa; on the east by

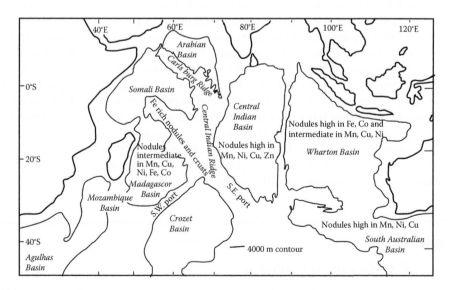

FIGURE 1.3
Indian Ocean showing different basins.

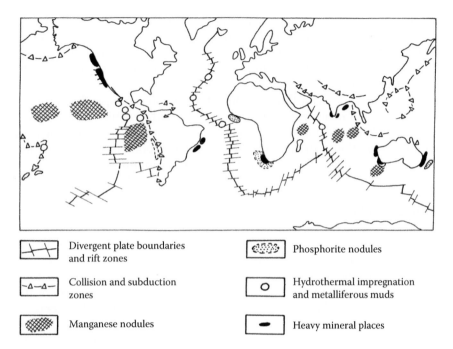

Divergent plate boundaries and rift zones

Collision and subduction zones

Manganese nodules

Phosphorite nodules

Hydrothermal impregnation and metalliferous muds

Heavy mineral places

FIGURE 1.4
Map of world oceans showing major tectonic zones and mineral occurrences.

the Malayan peninsula, the island of Sumatra, and the continent of Australia; and on the south by the icy continent of Antarctica. An arbitrary separation from the Atlantic Ocean is made at a longitude of 200°E off the Cape of Agulhas and from the Pacific at a longitude of 147°E off Tasmania. India has 2700 nautical miles of coast line, of which 80,000 sq. nautical miles is about 100 fathoms, 20 sq. nautical miles between 100 and 500 fathoms, and 30,000 sq. nautical miles is between 500 and 1000 fathoms.

The Indian Ocean is often considered as a tropical ocean, but actually it extends to latitude 700°S, from the Red Sea and Persian Gulf at 300°N. It has an average water depth of 12,700 ft (3890 m) and a volume of 70,086,000 cubic miles (292,131,000 cubic km). The maximum depth 24,442 ft (7450 m) is in the Java Trench. An idea of the depth zones of the Indian Ocean and its connected smaller seas, the Red Sea and the Persian Gulf, can be obtained from Tables 1.1 and 1.2.

TABLE 1.2

Depth Zones of the Indian Ocean (km²)

	Total Area	<200 m		200–1000 m		>1000 m	
		Area	%	Area	%	Area	%
All Oceans	105.567	7.909	7.49	4.669	4.42	4.630	4.38
Indian Ocean and Seas	21.613	0.889	4.10	0.632	2.92	0.786	3.64
Indian Ocean	21.461	0.705	3.57	0.575	2.69	0.766	3.58
Red Sea	0.132	0.055	41.45	0.057	43.06	0.020	4.92
Persia Gulf	0.069	0.069	All	–	–	–	–

Compiled from Different Sources.

Though the Indian Ocean is the only major ocean that does not extend from pole to pole, it is large enough that it extends from the Indian peninsula to Antarctica, spanning more than 10,000 km. Indian Ocean covers 73,426,500 sq km, one-seventh of the surface of the earth. The world's oldest and most densely populated countries fringe the northern part of the Indian Ocean rim.

Unlike the Atlantic and Pacific Oceans, the Indian Ocean is unique that it is closed from the north by the Asian continents, but open in the south. Further, the Indian Ocean is almost symmetrical about a north–south axis running down the length of the archipelago of Maldives. It is seen that if the Indian Ocean is folded about its axis, several megacities fold into one another, such as Dubai in UAE into Kolkata in India.

Similarly, Singapore falls along Mogadishu (Somalia) and Durban (South Africa) into Perth (Australia). This is being currently investigated into scientific aspects (http://geoarchitecture.file.2003.foldedocean) (Table 1.3).

Scientific investigation of the Indian Ocean dates back to the twentieth century. In the latter part of the nineteenth century, HMS *Challenger* and the German vessels *Gazella* and *Valdivia* provided a general understanding of the topography, which was later expanded by the expeditions of the *Sea Larke* and British cable ships. *Discovery IT* made extensive cruises beginning in 1929, in the high latitudes, and in the western part of the ocean. The oceanographic vessels *Willibroad, Snellius,* and *Dane II* worked in the open ocean and the *Ammireglio Magnaghi* in the Red Sea. The *Mabahiss* explored the northwestern part of the ocean during 1933–34; the Swedish *Albatross* surveyed the equatorial regions in 1948; and the Danish *Galathea* prospected in the deep waters in crossing from southern Africa to India and Indonesia in 1951.

The major activity in the Indian Ocean exploration and scientific work started during the International Indian Ocean Expedition (IIOE). Ships from different nations including Australia, Britain, Germany, France, India, Japan, Pakistan, Portugal, the United States, and the USSR (now CIS) participated in this expedition. Research activity continued, and during the monsoon, similar expeditions were organized with international cooperation. In addition, several ships have traversed the Indian Ocean to collect and study specific

TABLE 1.3

Continental Shelf Areas of the Indian Ocean
(Seabed <200 m/656 ft)

Ocean/Subsea	Km²	Miles²
East Africa	390,000	150,540
Arabian Sea	440,000	154,400
Bay of Bengal	610,000	235,460
Indonesia	130,000	50,180
Western Australia	380,000	146,680
South Australian Coast	260,000	100,360
Red Sea	180,000	69,480
Persian Gulf	240,000	92,640
Madagascar	210,000	81,060
Ocean Islands	200,000	77,200
Total Indian Ocean	3,000,000	1,158,000

Source: Bramwell, M. (ed). 1977. *The Mitchell Beazley Atlas of the Ocean.* Mitchell Beazley, London, 208 pp.

occurrences of various mineral deposits on the seafloor. The Ocean Drilling Programme (ODP), and its earlier version the Deep Sea Drilling Project (DSDP), have drilled several boreholes in the bed of the Indian Ocean. The study of such boreholes has enabled scientists to present a more complete history of the geological formation of the Indian Ocean.

Oceanic water plays an important role for India because in the classical sense, India forms a peninsula surrounded by the Arabian Sea in the west, the Bay of Bengal in the east, and the main Indian Ocean in the south. The Andaman and Nicobar group of islands and the Lakshadweep islands are surrounded by ocean on all sides. The Indian Ocean is the third largest water body in the world, next to the Pacific and the Atlantic Oceans. The Indian Ocean occupies an area of 73,600,000 km^2, constituting one-seventh of the earth surface, and represent 20% of the total ocean areas. Unlike the Pacific and the Atlantic Oceans, our Indian Ocean has a peculiar situation that it is not open on both ends. In the north, it is closed by the Asian continent. An arbitrary separation of Indian Ocean is made at longitude 20°E, off the Cape of Agulhas, and from Pacific Ocean at longitude 147°E, off Tasmania (Figure 1.1). The Indian Ocean is often considered to be a tropical ocean, but it actually extends to latitude 70°S and 30°N through the Red Sea and Persian Gulf. The Indian Ocean has an average water depth of 3800 m and volume of 292,131,000 km^3. The maximum depth of 7450 m is found in the Java trench. The Arabian Sea and the Bay of Bengal also have their special features. Mighty rivers such as the Indus with its major tributaries Sutlej, Beas, Ravi, Chenab, and Jhelam bring in an enormous amount of detritus material of the Himalayas through whose steep slopes and gorges they cut through. This enormous load is poured into the Arabian Sea at the mouth of the Indus river near the port of Karachi in Pakistan. The dispersal of this detritus material depositing at a very rapid rate has given a very important sediment piled feature in form of a "fan." This feature is called the Indus fan. Likewise in the Bay of Bengal, extending from the Sunderban delta region near Kolkata (Calcutta), a thick pile of sediments thousands of meters thick has been generated due to the deposition of detrital material brought by both Ganges and Brahmaputra and associated river systems. Again, here the material has been derived from the Himalayas. This mighty sequence of sediments extending thousands of kilometers from the mouth of Ganges to the equator and beyond forms the largest deep-sea fan called the deep-sea fan of Bay of Bengal. In many ways, the earth scientists have worked out a relationship between the history of the deep-sea fan of Bengal and the deep-sea fan of Indus from the period of the upliftment formation of Himalayas. The sediments of Bay of Bengal are predominately clay and silt, which are brought by the Ganges–Brahmaputra river system. These land-derived sediments extend in the south till one meets with the deep-sea calcareous sediments. The famous Ninetyeast Ridge does not act as a barrier for the sediments to be transported in the southern direction. In fact, the sediments become finer in nature as one moves south. Sediments from the deep-sea province, which really make up 80% of the ocean, have large part of their origins from biomass in the water column. In the Arabian Sea, overlying them within the nonbiogenic clays is the influence of Indus-derived sediment like the Bengal fan sediment quite near the equator. On the east coast of India, there are other large rivers such as the Mahanadi, Godavari, Krishna, and Cauvery, which bring in a lot of material from the Deccan plateau and deposit along the shores of the coast line. Some of them have attained great thickness and because of their organic content are potential reservoirs for hydrocarbon and natural gas, their gas hydrates, as are the Bengal and Indus fans. The exploration conducted by the Oil and Natural Gas Corporation (ONGC) and Indian Oil Limited have located gas reservoirs in the Godavari delta region. On the west coast, the major rivers, which flow into the Arabian Sea, are Narmada and Tapti. These rivers originate from central India and their important tributaries such as Chambal pour out a lot of material from central Indian plateau, such as sandstone and limestone into the Arabian Sea.

It was during the 1960s that the IIOE took place, which included ships from countries such as Australia, Britain, Germany, France, Japan, the United States, Russia, Portugal, Pakistan, and India. Our knowledge of the oceanographic, geological, geophysical parameters of the Indian Ocean got an impetus from this expedition. At present, India is paying enough attention to ocean science, and several research ships of the Indian government are operating around India to produce scientific information. Through these investigations, we have learned a lot about deep-sea pelagic sediments. Based on these studies, it is known that the Indian Ocean has three types of sediments.

Description of the sediments obtained from the Indian Ocean shows a large province of calcareous ooze sediments in the western region, a small red clay province in the east, and a circular diatomite zone in the middle. In addition to these sediments, a huge amount of land-derived terrigenous sediment derived from the continental and coastal areas predominate, as mentioned, with the detrital material deposited from the mouth of rivers Indus in the west forming the Indus fan in the Arabian Sea and like this the Ganges–Brahmaputra system depositing the largest deep-sea sediments called Bengal fan in the Bay of Bengal. If one attempts to quantify the distribution of sediments in the total Indian Ocean, one can generalize to say that the Bengal fan alone comprises almost 34% of volume of sediments in the Indian Ocean, the Indus fan comprises about 10%, the Pelagic sediments about 13.5%, and the African coastal basin 22.5% and Australian coastal basin about 2%, and other smaller units make up to 17.5%. Full information on the Indian Ocean's geology and sediment pattern is available in Narin and Stehli (1982).

1.3 Offshore Bed Rock Mineral Deposits

1.3.1 Geological Regime and Structure and Relief of the Indian Ocean

The profile from the shoreline to seafloor indicates two major physiographic regimes: continental margin, which represents the submerged edges of the continental masses, and the deep ocean basin. These physiographic contrasts between the landmass and oceanic basin are governed by the geological events that have occurred below them (i.e., within the crustal and mantle layers), and huge loads of varying types of sediments have been deposited over them. The mineral resources in the various subsea physiographic provinces are the product of geological processes. Mineral deposits are present in all parts of the ocean. Minerals containing zinc, copper, nickel, and cobalt occur in the mid-oceanic ridge system and on the oceanic plain, away from the continents, as well as deposits of aggregates and heavy minerals in the shallower inshore waters. Therefore, a basic review of the geological history of the Indian Ocean has been presented below.

Magnetic studies by McKenzie and Slater (1973) provided information on the structural development pattern and the age of the Indian Ocean, and the deep boreholes by the *Glomar Challenger* (of the National Science Foundation, 1974–75, 1988–90) added more recent information to our knowledge of the Indian Ocean. One may refer to Narin and Stehli (1982) and Roonwal (1986) for earlier summaries. Based on this information, rocks of the Jurassic to Chalk, 140 million years old (ma), and crust accretionary material produced through the basaltic lava flows form the continental plates, and the sedimentary sequence is thin over these basaltic flows. The deep cores obtained through the DSDP borings indicate them to be between 154 and 725 ma. A part of the central Indian Ocean Ridge is an active ridge, and therefore the sedimentary cover is negligible here. The maximum sediment sequence is exposed in the Arabian Sea and the Bay of Bengal regions where the middle thickness of the Miocene sediments alone is about 800 m. The oldest sediments over the floor basalt are in the Wharton Basin, northwest Australia. These belong to early Chalk or to the Tithon formation. The oldest magnetic anomalies are 75 million years (upper Chalk) old. However, in most of the drill bores of the *Glomar Challenger*, the most common sediments were formed during Paleocene to Oligocene age. Therefore, by Oligocene time, the Indian Peninsula may have moved away from South Africa and occupied its present position. The Owen Fault and the Ninetyeast Ridge mark the relief of the main horizontal movement. Since then, the active centers have contorted and twisted to be now on the Central Indian Ridge and its extension—the Carlsberg Ridge in the north, and the both arms of the Central Indian Ridge in the south.

The main pattern of the separation took place around 35 million years ago, with a movement of nearly 1.2 cm/year in the Carlsberg Ridge and 2.3 cm/year in the southern limits of the ridge (Laughton et al., 1970). The southwest setting shows a younger age and continental drifting at 1 cm/year rate. The details of the physiographic limits showing the structural lineaments of the ocean are graphically illustrated in the map of Heezen and Tharp (1966) and as well as the physiographic map prepared by UNESCO (Udinstov, 1975).

Two of the most conspicuous structural lineaments of the Indian Ocean are the Mid-Oceanic Ridge, which has the configuration of an inverted "Y" branching at a point northeast of Mauritius and the Ninetyeast Ridge (Figure 1.2). The Mid-Oceanic Ridge, also known as the Central Indian Ridge, begins as "Shaba ridge" from the Gulf of Aden and then turns south to form the Carlsberg Ridge. The main Central Indian Ridge merges with the Rodrigues fracture zone on the Mauritius plateau in the southeast and southwest branches (Fischer et al. 1971). The ridge is cut off at several places by many transfer faults, thereby making several morphological features through deep grabens. In the central zone of the ridge, bottom photographs reveal a complex morphology in parts with sheer rock outcrops. The rocks are generally basaltic in nature. In several places, however, ultramafic and other eruptive rocks were obtained, and strong hydrothermal flows identified (Herzig and Pluger, 1988, Pluger et al. 1990).

West of the Mid-Indian Ocean Ridge lies the large island of Madagascar, and the Seychelles Bank, which remained in their place during the drifting of the Indian peninsular shield, in the process producing several deep basins—the Madagascar Basin, the Natal Basin, Somali Basin, and the Mascarne Basin. The granitic Seychelles Islands and associated bands lie at the northern end of the Mascarne Ridge, showing up as islands and reefs, and on the southern side are the Mauritius islands, which are typical of volcanic origin.

The Somali Basin and the adjoining basins were produced by gliding of one of the major western lineaments of the ridge. The base of the floor is Upper Cretaceous, over which lie hundreds of meters of sediment, deposited during the course of time. In the Natal Basin, also in the area near the continent, are thick sequences of sediments, for instance, in the Mozambique Basin 400 m of Palaeocene sediments are to be found.

The Madagascar basin has a basement of early Tertiary basalts, with minimal sedimentary cover. Between the two southern arms of the Indian Ocean Ridge lies the Crozet Basin, about which little is known. The northern part of the Indian Ocean divides into the formations of the Arabian Sea and the Bay of Bengal, through which the Indian continental stream water flows, depositing a thick sedimentary pile. According to seismic measurements, the thickness of the sediments varies between 2.8 and 8 km (ONGC, India 1980).

The two sediments are separated by the Chagos-Laccadive Ridge, which is of volcanic structure surrounded by coral reef.

East of the Central Indian Ridge lies the Central Indian Basin, the largest basin in the Indian Ocean and it has an oceanic crust under a few hundred meters of sedimentary cover of Tertiary to recent sediments. In the northern part, these basin sediments come under the influence of the Ganges Delta sediments. The Central Indian Basin is bordered in the east by the Ninetyeast Ridge. The Central Indian Basin contains sediments similar to those described from the other basin on the western side of the Mid-Indian Ocean Ridge. Clearly, it relates mainly to a fracture zone with a graben-horst tectonic pattern (Marine Geology, NGRI). On the eastern side of the Ninetyeast Ridge, the basin continues in a similar fashion to the main Indian Ocean Basin and has been named Keeling Basin in the Sundra Graben within the subduction zone. West of Australia lies smaller deep-sea basins, which have been defined through the plateau or submarine range of mountain chains. The detailed structural setup of the area is complicated and in this area occur the oldest sediments of the Indian Ocean in the Wharton Basin and the continental part represented by Broken ridge.

1.4 Sediment Pattern

Earlier descriptions of the sediments obtained from the Indian Ocean (Schott, 1935) indicated a large province of globegerina sediments in the western region, a small red clay province in the east, and a circular diatomite zone in the middle. In addition, in these sediments are patches of terrigeneous sediments derived from the continental and coastal areas of the north, where they predominate (Figure 1.5).

The investigations during the IIOE and successive studies have added to our knowledge of the sediment pattern in the Indian Ocean. The deep-sea or pelagic sediments are known through seismic measurements and through drilling by *Glomar Challenger* (DSDP), and *Glomar Explorer* (ODP). The map prepared by Bezrukov (1973) shows the different sediment types and their strong differences. The terrigeneous sedimentary cover stretches through the following: the eastern coast of Africa into the Arabian Sea, the Bay of Bengal, and then along the Indonesian archipelago. Shallow-water biogene sediments are found along some coasts and on the high but inactive ridges, such as the Chagos-Laccadive Ridge and Mascarne Ridge. The major part of the ocean bed is covered by pelagic sediments. The ridges and the basins have an average water depth of 4700 m, containing calcareous sediments (primarily foraminifera rich and neno-sediments). In the deeper ocean basin are found clayey sediments, especially north of the equator and up to 15°S. In the southern equatorial zone, in between are found siliceous sediments, dominantly diatom sediments. In the southern clayey sediment bar lies a second siliceous sediment ring, again dominated by diatom sediments. The pelagic calcareous, siliceous, and clayey sediments have a low thickness, generally around 100 m. There are three main types of sediments, with an average rate of sedimentation around of 1 mm/1000 year. Estimations of sediment rate, measured from deep borings, indicate that in the Indian Ocean, as in the other world oceans, the actual rate depends on zones of local sediment rate. Also the shelf sediments are not mainly of Holocene age. The total distribution of sediments in the Indian Ocean is summarized in Table 1.4.

From the above, it can be seen that nearly 50% of the sediment volume in the Indian Ocean lies in the Arabian Sea and the Bay of Bengal. The sediment of the Arabian Sea

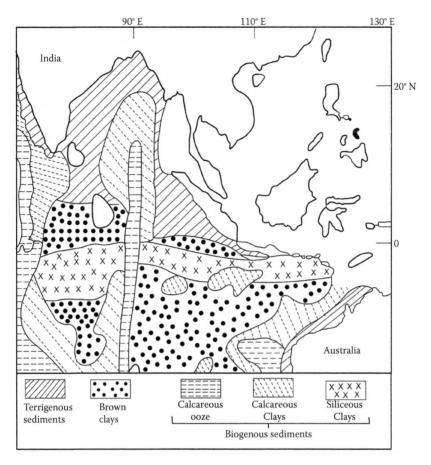

FIGURE 1.5
Indian Ocean showing main sediments types. (Based on Roonwal, G.S., 2002. Cooperation in non-living resources—An Indian prospective. In: P.V. Rao (ed). *India and the Indian Ocean*, South Asian Publisher, New Delhi, 239, pp. 88–99.)

consists mainly of bands of the clayey sediments of the Indus. A strong influence of wind-blown (aeoline) sediments is seen in the northwest basin area near Oman.

The sediments of the Bay of Bengal are predominantly clays and silts, which are brought by the Ganges and its tributaries and other major rivers. This extends in the south till one meets with the pelagic calcareous sediments. The Ninetyeast Ridge does not act as

TABLE 1.4

Distribution of Sediments in the Indian Ocean

Area	% Volume
Indus Delta	9.9
Ganges Delta	34.2
African Coastal Basin	22.6
Australian Coastal Basin	2.1
Pelagic Sediments	13.4
Others	17.6

a barrier for the sediments to be transported and dispersed due south. The sediments become finer in nature, as one moves south. A large proportion of the clastic component in the sediments is reworked through turbidities and turbidity currents. Drill bores of 211 sites during the DSDP, for example, shows 773 m thick middle Miocene to recent sediments with well-preserved turbidities in alternating bands of calcareous sediments, contained within them. The sediments from the pelagic provinces, which make up to 80% of the Indian Ocean, to a large part, have their origin from biomass in the water column. Overlying them within the nonbiogenic clays is the influence of the Indus River falling into the ocean; it is seen clearly up to the equator, especially marked by the decrease in the quantity of such clay minerals as illite and chlorite (Kolla and Kidd, 1982; Roonwal and Srivastava, 1991). The influence of the volcanic material is seen through the presence of varying quantities of montmorillonite, which in part may also be of secondary origin from the detrital material (maximum for the Deccan basalt of the Indian peninsula). More than half the quantity of montmorilonite is observed in the eastern part of Mascamesea, which could be due to the above reason. The zeolite (phillipsite) is present as the weathering product of basaltic volcanic material in more than 10% of the surface sediments of the abyssal plains of the Central Indian Ocean (Kolla and Kidd, 1982). There is the influence of undersea volcanism as well as land volcanism in the sediments. The land volcanic material appears also to have been brought by wind, as can be deciphered by the variation in the type and percentage of the zeolites in the sediments. Different authors have different opinions about the presence of the wind-borne detrital material in the pelagic sediments within the abyssal plain. The largest influence is in the northwestern part of the Indian Ocean, which comes just outside the main area of sediments deposited by the Indus River (Roonwal, 1986, IOCINDIO-II). More details can be seen in Narin and Stehli (1982).

1.5 Exclusive Economic Zone

The EEZ of India extends 200 nautical miles from the shoreline and covers an area of two million sq. km. Although during the IIOE, data from the seas around India was collected, the real objective was to investigate the EEZ, which was declared in 1987. The overall scientific objective of this interest is to understand the oceanographic environment and assess the living and nonliving resources surrounding the main land. Consequently, research and development projects have been defined to achieve this goal. This includes coverage of the EEZ in respect of physical, chemical, biological, geological, and geophysical data collection and synthesis. Cruises in the zone have already resulted in collection of data from more than 16,000 stations (NIO annual report 1988–1989). This is, however, not to suggest that the EEZ has been fully surveyed, as large areas still remain to be investigated.

The oceanographic features around the Indian mainland witness seasonal as well as annual change during the monsoons. There is both national and international interest in investigating this phenomenon within oceanographic parameters.

These are essentially seaward equivalent of onshore deposits and include oil and gas, coal, metalliferous deposits, sulfur, and evaporate. They can extend offshore for hundreds of kilometers and down to depths of thousands of meters below the sediment–water interface. For the most part, deposits of this type are exploited by modified onshore type operations such as the use of various types of drilling platforms in the case of hydrocarbons. Drilling and solution mining are used for sulfur in areas such as the Gulf of

Mexico and could considerably be used for offshore extraction of evaporities particularly potash. Subsea mining of coal from shore-developed operations has been underway for many years in Japan, the UK, and Canada. Similarly, metalliferous deposits have been mined offshore in many parts of the world. Also much of the world oil and gas production, which is located offshore, occurs in the EEZ and constitutes significance national resources.

1.6 Shallow Oceanic Mineral Deposits

It is toward the near shore zone that the greatest efforts in India could probably be directed where sea-level changes, environments, tectonics, and resources evaluation are achievable. In addition to the oil and gas, a number of marine commodities, notably the coastal zone as a resource in its own right, is significant because mineral sands contain rutile, zircon, ilmenite, magnetite, and monazite. Construction materials such as gravel, sand, and carbonates, lastly the phosphorites (sedimentary phosphate deposits), some of which are being explored from the seafloor at the present. These resources will be considered in detail later in the following chapters.

1.7 Deep Oceanic Mineral Deposits

Deep-sea mineral deposits are found in open ocean down to abyssal depths usually far from the land where chemical and biochemical sediments are not diluted by input of land-derived or terrigenous material. No deep-sea deposits have been commercially exploited to date. There have been various attempts to commercially evaluate them and it is evident from these that development and mining cost will be enormous. Up to now, the main interest has been focused on pollymetallic manganese modules, which are found in many parts of the deep ocean at abyssal depths, and of late, interest has grown in the massive sulfides found as cluster of chimneys at several locations in the mid-oceanic ridges, as well as the cobalt-rich crusts found on many of the ancient volcanic seamounts in the areas of upwelling. The commercial interest in Mn nodules is not in their Mn but rather in the Ni, Cu, and Co contents. In the areas of the northern Pacific, the $Cu + Ni + Co$ content in nodules is about 3% and Mn around 25% to 30%. It is in this Pacific area that much of the commercial interest has been focused by countries like Japan, China, former USSR, the United States, France, and Germany, which have carried out prospecting, exploration, and even claim some areas for the future. In the Indian Ocean, Mn nodules occur in several locations, which are defined in each oceanographic basin, details of which have been discussed later. The massive sulfides have recently (1994) been located around the Triple Junction of Mid-Indian Ocean Ridge. There is a strong likelihood of massive sulfide deposits also occurring in the Andaman Sea, details of which are given in Chapter 6. Because cobalt-rich crust is found in the open ocean mainly on the upper slopes, marginal plateaus and seamounts at depths of about 1000–2000 m rather than at depths of 4000–6000 m that nodules characteristically occur, this relatively shallower water depth may make them commercially more attractive. But

TABLE 1.5

Nonliving Resources of the Ocean

- Polymetallic deposits forming hydrothermal beds, mounds, and chimneys on the seabed, and stock work in the underlying rocks.
- Ferromanganese nodules and crusts.
- Phosphorites found on seamounts, submarine plateau continental shelves, and slopes
- Evaporates such as those of the Mediterranean Sea
- Heavy minerals such as tin, platinum, diamonds, gold, rutile, monazite, etc.
- Hydrocarbons formed from rapid maturation of organic-rich sediments by hydrothermal fluids.
- Thermal energy that can be tapped from hot hydrothermal vents, from high temperature gradients in rock, and by using the ocean's own gradient.

unlike Mn-nodules, the crusts are likely to be irregular in their occurrence, difficult to extract from the seamounts.

The occurrence of sulfide-bearing minerals and sediments in association with "black smokers" has been a subject of a great deal of scientific attention in recent years. These deposits were first discovered in 1999 on the East Pacific Rise where hydrothermal vents form a series of "chimneys" along parts of the spreading ridges. Subsequently they have been found in the vicinity of other spreading centers. The metalliferous sediments contain high concentration of Cu, Zn, Pb, Si, and Fe sulfides in places. Some of this is precipitated, but the hot dense brines exalted from the vents can form metalliferous muds that are many kilometers away from the actual vents. Metalliferous muds and brines of the Red Sea are also a type of sulfide resource but with somewhat different origin. The hot metal-rich brines are derived from the circulation of heated groundwater evaporities of the Red Sea graben, resulting in sediments and brines rich in metals such as Cu, Zn, and silver (Table 1.5).

India is engaged in marine mineral exploration. Serious and directed efforts have been made in respect of the evaluation of Mn-nodules in the Central Indian Basin. In the absence of backup information on environment related to mining operations, and a complete absence of deep-sea mining technology within the country, even test mining cannot be conducted as both these are important parameters. The recent mining test performed in collaboration with Germany is to be noted.

References

Amann, H., 1982. Technological trends in ocean mining. *Philos Trans R Soc London*, 307: 377–403.

Antrim, C. L., 2005. What was old is new again: Economic potential of deep ocean minerals the second time around. In: *Background Paper of the Center for Leadership in Global Diplomacy: Oceans 2005 Conference*, Washington, DC, p. 8.

Bezrukov, P. L., 1973. Principles scientific results of the 54th Cruize of RV Vityaz in the Indian Ocean and Pacific Ocean. *Oceanology*, 13: 761–766.

Bramwell, M. (ed). 1977. *The Mitchell Beazley Atlas of the Oceans*. Mitchel Beazley, London, p. 208.

Cruickshank, M. J., 1998. Law of the Sea and mineral development. In: E. Mann Borgese (ed). *Ocean Year Book*, 13. University of Chicago Press, Chicago, pp. 80–106.

Fischer, R. L., Sclater, J. G., and McKenzie, D. P., 1971. The evolution of the Central Indian Ridge, Western Indian Ocean. *Bull Geol Soc Am.*, 82: 553–562.

Heezen, B. C. and Tharp, M., 1966. Physiography of the Indian Ocean. *Philos Trans R Soc Lond A*, 259: 137–149.

Hein, J. R., McIntyre, B. R., and Piper, D. Z., 2005. Marine mineral resources of Pacific Islands - A review of the Exclusive Economic Zones of islands of U.S. affiliation, excluding the state of Hawaii: U.S. Geological Survey, Circular 1286, p. 62.

Herzig, P. and Pluger, W. L., 1988. Exploration for hydrothermal activity near the Rodriguez triple junction, Indian Ocean. *Canadian Mineralogy*, 26: 721–736.

Hsu, K. J. and Thiede, J. (eds). 1992. *Use and Misuse of the Seafloor*. Dahlem Konferenzen. Wiley, New York, p. 440.

Intergovermental Oceanographic Commission of the UNESCO. 1982 onwards.

International Indian Ocean Expedition of the UNISCO. 1959–1965.

Kolla, V. and Kidd, R. B., 1982. Sedimentation and sedimentary processes in the Indian Ocean. In: AEM Narin and FG Stehli (eds). *The Ocean Basins and Margins, vol. 6. The Indian Ocean*. Plenum, New York, pp. 1–50.

Laughton, A.S., Matthews, D.H., and Fischer, R.L., 1970. The structure of the Indian Ocean. In: Maxwell A.E. (ed). The Sea, vol. 4, Wiley Interscience, New York, pp. 543–586.

Lenoble, J.P., Augris, C., Cambon, R., and Saget, P., 1995. *Marine Mineral Occurrences and Deposits of the Exclusive Economic Zones*. MARMIN: A Database: Edition, IFREMER, Brest, France, 274 pp.

Mann Borgese, E., 1992. Ocean mining and its future of world order. In: K. J. Hsu, and J. Thiede (eds). *Use and Misuse of the Seafloor*. Dahlem Konferenzen. Wiley, New York, pp. 117–126, 440 pp.

McKenzie, D. P. and Sclater, J. G., 1973. The evolution of the Indian Ocean. *Sci Am.*, 228(5): 62–72.

Narin, A. E. M., and Stehli, F.G. (eds), 1982. *The Ocean Basins and Margins, vol 6. The Indian Ocean*. Plenum, New York, p 540.

NIO Annual Reports 1988 onwards.

ONGC, India 1980 onwards.

Pluger, W. L., Herzig, P. M., Becker, K. P., Deismann, G., Schops, D., Langer, J., Jensich, A., Ladage, S., Richnow, H. H., Schulzee, T., and Michaelis, W., 1990. Discovery of Hydrothermal Field at the Central Indian Ridge. *Marine Mining*, 9: 73–86.

Rona, P.A., 2008. The changing vision of marine minerals. *Ore Geology Reviews*, 33: 618–666. doi:10.1016/j.oregeorev.2007.03.006.

Roonwal, G. S., 1986. *The Indian Ocean: Exploitable Mineral and Petroleum Resources*. Springer-Verlag, Heidelberg, Vol. XVI, p. 198.

Roonwal, G. S. and Srivastava, S. K., 1991. Clay mineralogy of the pelagic sediments along a west-east transect in the Indian Ocean. *J Geol Soc India, Bangalore*, 38: 37–54.

Roonwal, G.S., 2002. Cooperation in non-living resources—An Indian prospective. In: P.V. Rao (ed). *India and the Indian ocean*. South Asian Publishers, New Delhi, pp. 88–99, 239 pp.

Schott, G., 1935. *Geographie des Indischen und Stillen Ozeans*. Boysen, Hamburg, p. 413.

Scott, S.D. and committee members, 2008. Mineral deposits in the sea: Second report of the ECOR panel on marine mining, 34pp.

Thiede, J. (ed). 1983. Whither the oceanic governance. Report III. International workshop on marine resources IUGS.

Udinstov, G. B., 1975. *Geological-geophysical atlas of the Indian Ocean*. Acad Sci USSR, Moscow, p. 152.

2

Ores in the Deep Sea: Manganese Nodules

2.1 General

The first large-scale scientific exploration of the sea was initiated by the Royal Society of London in 1870 and in December 1872 the ship HMS *Challenger* specially equipped for deep-sea research left England for a voyage, which lasted three years and covered all the major oceans. The scientists under the direction of Sir C. W. Thomson lifted with their dredges a multitude of living creatures, many of them unknown before, as well as organic and mineral deposits from the seafloor. Among their finds was the first specimen of ferromanganese nodules. In 1891, the participants of the expedition John Murray and A. Renard (1891) presented a detailed report of these samples. The nodules had come from all oceans. The results of the British expedition prompted other nations to promote deep-sea expeditions of their own. In 1906, the U.S. research vessel *Albatross* collected similar samples of manganese nodules and about the same time the German ship *Valdivia* returned from worldwide expedition (see Glasby, 1977 for details).

Not much oceanographic research was undertaken until after the Second World War. One of the results of these expeditions was that for the first time greater quantities of manganese nodules became available and so it was possible to start a wider scientific study of this new type of ore. Nevertheless, it was not before the end of the 1950s that detailed results of studies were published concerning the occurrence, structure, and formation of manganese nodules. It is the book *The Mineral Resources of the Sea* (Mero, 1965) to which we owe the present-day interest in the economic possibilities of this new type of ore. This publication caused oceanographic institutes as well as industrial concerns to treat manganese nodules more seriously as a possible source of future supplies of some nonferrous metals like nickel, copper, and cobalt.

Ocean floor nodules represent the largest group of ferromanganese deposits. Manganese nodules have been reported from all the oceans, however, nodules occurring from the Pacific and Indian Ocean are considered to be economic interest. The collection of nodules from the Seychelles basin of the Indian Ocean during the 86th cruise of R.V. *Gaveshani* of Goa-based National Institute of Oceanography in January 1981, was a step in this direction. In the Pacific Ocean, best mining sites have been identified in the Central Pacific Ocean and South East Pacific Ocean (Cook Islands). In the Indian Ocean, best quality nodules occur in the Central Indian Ocean Basin. Therefore, as it was expected the central part of the Indian Ocean has shown large deposits of ferromanganese nodules. It is in this region that the Central Indian Basins that India has identified areas for lease for mining.

2.2 What are Mn-Nodules?

Manganese nodules primarily comprise of iron and manganese hydroxides. They are generally potato-like nodular deposits and have grown around a nucleus or "seed" of such material as a weathered basaltic grain, clay, or phosphatic material, dead animal remains or even the iron pieces of sunken ships and bomb shells, indicating that they are of recent origin.

Manganese nodules are earthy black in color when wet and brown when dry (Figure 2.1). However, the color might vary from black to tan. Hardness of the nodules is variable, ranges from 1 to 4 on Moh scale of hardness. That is, they can be so soft as to be broken by a light pressure between two figures, or hard enough to withstand light taps with a hammer. The normal hardness is however 2–3 Mohs. It is observed that nodules, which have about 5% of $CaCO_3$ are hard and, those with less than 2% $CaCO_3$ are soft and therefore friable. Clays present in nodules also control the hardness. Since manganese nodules are porous, they are light with specific gravity of 2–3. The shape of nodules is variables. They are often spheroids to oblate, discoidal, or prolate. Generally, they are asymmetric. True symmetrical shapes in nodules is rare. Such spherical shapes are seen in small nodules having diameter less than 2 cm. They are found as single as well as combination of two or three joined together. They occur scattered on the sediments of the seafloor.

The age of a nodule is generally defined to be the time when the first oxide layer beings to form. The age or rate of accretion of manganese nodules can be determined by dating the nuclei or "seed" around which the ferromanganese oxide layers accumulate, or by assessing the age of differences among the various successive layers of these oxides within the nodule. The "seed" or nucleus dating gives only a maximum age for the nodule. Radioisotopes have been chosen to make accurate determination of dating of nodule. By 230th method, some nodules from the Indian Ocean have been given growth rate around 2.8 mm/10^6 years (Somayajulu et al., 1971 for details).

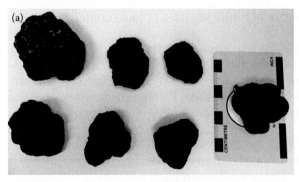

FIGURE 2.1
A close look of manganese nodules (nodules supplied by National Institute of Ocean Technology, Chennai). (a) Assortment of manganese nodules from the Central Indian Ocean Basin, and (b) Close look at manganese nodules from the Central Indian Ocean Basin.

2.3 Distribution of Manganese Nodules

Manganese nodules from the seabed were first collected by the ship *Challenger* of England at a depth of about 300km southwest of the Canary Islands in the Atlantic Ocean. This occurred during the period 1872–1876.

Subsequent surveys and research have proven that manganese nodules are distributed in almost all deep seabed around the world. However, it was not until the late twentieth century that the distribution areas of economically feasible manganese nodules became clear.

Once it became clear that the total grade of nickel, copper, and cobalt of the manganese nodules, distributed in the seabed of the so called C–C zone (the Clarion–Clippertone fracture zone) southeast of Hawaii in the Pacific Ocean, is about 2.5%, large enterprises from the United States, Germany, and other countries actively searched for nodules from the 1960s to the early 1970s. As the surveys, and the research and development of manganese nodules required highly advanced technology, as well as a huge amount of funds, large enterprises from the United States, Germany, Japan, and others organized an international consortium around 1975 to conduct joint surveys and research and development. As a result, INCO, US Steel, KENCOTT, and others conducted mining experiments to exploit manganese nodules on the deep seabed and succeeded in recovering manganese nodules at a depth of 5000 m.

2.4 Deep-Sea Mineral Resources

Manganese nodules comprise hydrated iron and manganese oxides. They contain nickel, copper, and cobalt, a spherical or elliptical form with nodular diameters ranging from 1 to 15 cm. Abundant manganese nodules in the Central Indian Ocean Basin occur at a water depth of 4000–6000 m on the seafloor.

Manganese nodules occur in various shapes: hamburger steak, ball and knot shape; 1 to 15 cm in diameter and dark brown to black in color. They contain iron and manganese hydroxides as the main components and metals such as copper, nickel, cobalt, molybdenum, and titanium. Generally, a dark brown manganese nodule is richer in iron. A friable manganese nodule shaped like a knot with a smooth surface often contains lower grade nickel and copper. A black manganese nodule, like a hamburger steak with a rough surface, generally contains higher grade nickel and copper. The manganese nodules discovered at about a depth of 5,500 meters in the EEZ of the Cook Islands look like tennis balls and contain relatively high grades of cobalt compared with those of the C–C zone.

Most of the manganese nodules in the C–C zone are distributed on the seabed at a depth of 4000 to 6000 meters and contain relatively high grades of nickel and copper. At a depth of 1000 to 2500 meters, on both seamounts and sea areas near the land, manganese nodules are also found. Manganese nodules distributed on seamounts, compared with those on the seabed at 4000 to 6000 meters in depth, often contain higher grades of cobalt. The metal grade of manganese nodules distributed at a relatively shallow depth and near land is generally low. Moreover, ultrasmall manganese particles, which are called micronodules, are distributed in sea water and/or on the seabed.

Many manganese nodules form in a concentric circle around a nucleus of a rock chip, a dead organism, or a shark's tooth. The manganese nodules are distributed on pelagic sediment; some are concentrated and some are scattered, while others are covered with sediment. At depths where manganese nodules are distributed, mega benthic fauna such as fish, shrimp, and other organisms live; in addition, in the sediment of the seabed, meio benthos, which are too small to be photographed by a deep-sea camera, live. There is other mysterious benthic community; however, we have not yet enough knowledge, data, and information on them. In recent years, the biocenoses of the seabed have received attention. It is now necessary to start collecting basic data and information about the biocenose so that the effects will be minimal when commercial mining of manganese nodules begins. For such purposes, the United States, Japan, Germany, India, Poland, Korea, China, and others have been conducting marine environmental surveys, sampling, and research. Their research results were demonstrated at the International Symposium on Environmental Studies for Deep-Sea Mining, held in Tokyo from November 21 to 22, 1997. The symposium was organized by the Metal Mining Agency of Japan and supported by the International Seabed Authority and the Advanced Marine Science and Technology Society. The symposium was very informative and fruitful in understanding the present activities of marine environmental research of each country.

2.5 Growth Pattern, Mineralogy, and Geochemistry

Manganese nodules exhibit several types of surface textures. These textures can be described as smooth, gritty, goose bumps, pisolite, and knobby. In order to study their growth pattern and internal structures, generally a cross-section of a nodule is taken. A nodule is cut into two halves. The cut part of one half is polished to study various internal growth patterns and mineralogy by microscopy. Such studies also help to understand the distribution of elements Mn, Fe, Cu, and Ni in various ring zones with the help of electron microanalyses. The microscopic study of polished sections of nodules show clearly that they grow around a nucleus or "seed" as very fine concentric layers, often showing colloform to columnar pattern in a quite rhythmic pattern. Several minor features are also observed, which the suggest origin of growth of nodules under fluid conditions. The nodules are amorphous in nature. Therefore, no well-defined mineral phases are available. The most common minerals have been identified as Mn-rich variety todorokite (also called 10 Å manganite), very poor in Fe, which is the main manganese phase. Consequently, the metal supply for growth is based on Mn^{2+}, Ni^{2+}, and Cu^{2+}. The manganese-poor variety is 7 Å manganate (also called delta MnO_2), which is identical with birnessite. Next to manganese, iron is the most important mineral-building element, which appears rarely as amorphous FeOOH, but mainly as goethite. Its share in the nodule can be up to about 35 wt% (compared to manganese, which is 50 wt%). Other nodule components are minerals coming from volcanic seamounts in the oceans such as olivine, pyroxene, and zeolite; the minerals left over in the seafloor sediments, rutile, anatase; and some later formed minerals such as zeolite, phosphates, and clay minerals. During recent years, a number of detailed analyses of manganese nodules have become available. As a result, a number of other metals have been detected in them, such as Ag, Au, Bi, V, Cd, Ye, La, Sr, Ti, Sn, W, Zn, and so on. However, economic interest is concentrated on the contents of nickel and copper and then on cobalt and manganese.

The occurrence of metal-rich nodule in the oceans between depths of 3000–5000 m and far away from the continents reveals that nodules occur generally in areas with low average sedimentation rates (<1–3 mm/1000 years). Variations in the sediment accumulation rate can occur due to, for example, the activity of deep bottom currents, which on their part are influenced by the regional, topographic situation in the pelagic (deep sea) areas very fine-grained siliceous ooze to siliceous clays are dominant, and in such a situation the bottom is sufficiently below the carbonate compensation depth (CCD the depth below which no carbonate precipitates). This reflects that several factors are responsible for the growth of nodules. Also, aspects such as the speed with which the manganese nodules grow have been worked out by using radioactive isotopes. Calculations show that their growth can be between 1 mm/1000 years to 1 mm/1 million years, which means that, on an average, only a single layer of atoms a day is deposited (see Somayajulu for details).

2.6 Genesis of Manganese Nodules and Manganese Crusts (Tables 2.1 and 2.2)

Sorem (1973) presented some general views on the formation of manganese nodules, which by and large are accepted as correct:

- The nodules grow very slowly on the ocean bed through accretion of Fe–Mn hydroxides around a nucleus.
- The nodules are distributed over large areas of the ocean floor.

TABLE 2.1

Schematic Representation of the Possible Origins of the Elements, and the Potential Areas of Formation of Manganese Nodule Deposits

A. Origin of the elements (Mn, Fe, Co, Ni, Cu, Zn, etc.)	B. Potential areas of formation of manganese nodules and manganese crusts
1. Terrestrial weathering. Transportation by river water and seawater.	1. Regions with Fe–Mn-rich submarine igneous rock; high Fe and Mn supply in interstitial water due to submarine weathering.
2. Submarine weathering. Transport of dissolved elements by interstitial water.	2. Regions with increased accumulation rates of organic material. Fe and Mn are bound in organometallic compounds.
3. Submarine volcanism.	3. Regions with submarine volcanic activity; high Fe and Mn supply in hydrothermal solutions and volcanic exhalations.
4. Transport by hydrothermal solutions or by volcanic exhalations or by thermal mobilization.	4. Regions with low sedimentation rates. Accumulation of Fe and Mn from the seawater and interstitial water over a long period of time.
5. Cosmic material (cosmic spherules)	5. Regions with increased sedimentation rates of cosmic material.

Source: After Bonatti E., Kraemer, and Rydell H., 1972. *Ferromanganese Deposits on the Ocean Floor.* IDOE, Nat. Sci. Foundation, Washington, DC, pp. 149–902; Horn D. R., Horn B. M., and Delach M. N., 1973. *The Origin and the Distribution of Manganese Nodules in the Pacific and Prospects for Exploration.* Valdivia Manganese Exploration Group and Hawaii Institute of Geophysics, Office of the International Decade of Ocean Exploration, Washington, DC, pp. 77–83; Bonatti, E., 1982. *Hydrothermal Processes at Sea Floor Spreading Centres.* Plenum, New York, pp. 491–502.

TABLE 2.2

The Most Important Mineral Phases of Manganese Nodules and Their Correlation with the Elements Mn, Ni, Cu, Zn or Fe, Co

Reflectivity	Mineral Types	Element Correlation	
(520 nm)	7 Å, Manganite delta, MnO_2	(Na, Ca) $Mn_7O_{14} \cdot 2H_2O$	
−16%	Manganous manganite		Mn (Ni Cu Zn)
	10 Å, Manganite todorokite	(Mn, Ca) $Mn_3O_9 \cdot 2H_2O$	
	(Woodruffite, nsutite, rancieite, psilomelane) (pyrolusite) (vernadite)		
	Amorphous Mn–Fe oxides		
−8%	Amorphous Fe oxides		Fe (Co)
	Fe–Mn clay minerals		

Source: After Sorem R. K., and Foster A. R., 1972. *Ferromanganese Deposits on the Ocean Floor.* IDOE, Natural Science Foundation, Washington, DC, pp. 167–179.

- The chemical composition of the nodules, in particular their contents of the elements Ni, Cu, and Co, indicates a clear dependence on the geographical location and on the geological conditions and sediment facies at the ocean floor.

Sorem states that the nodules seen in those parts of the ocean floor, which have been covered in detail by photographic surveys, are approximately uniform in size but judging by the conditions, which we have discovered in the small areas investigated by us, this view must to be revised some extent.

The possible origins of the elements Mn, Fe, Co, Ni, Cu, Zn, etc. and the potential areas in which they were formed are shown in diagrammatic form in Figure 2.2.

According to Sorem (1973), there is still no general agreement on the following questions regarding the genesis of manganese nodules:

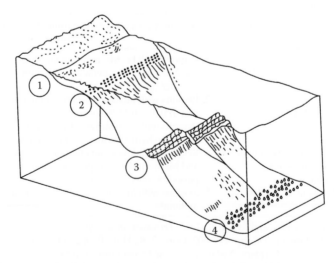

FIGURE 2.2
Location of Marine Minerals According to Topography of the Seafloor; (1) Heavy mineral placers; (2) Phosphorite; (3) Hydrothermal metalliferous impregnation and deposits; (4) Manganese nodules.

- Where do the nodules form in relation to the seawater/deep-sea sediment inter-face, that is, do the nodules grow (a) while they are lying on the ocean floor, (b) while they are embedded in the sediment, or (c) in both cases?
- Are the main and trace elements supplied mainly from the (a) seawater, (b) pore water, or preferentially from (c) residual submarine hydrothermal fluids?
- Do the nodules grow mainly as a result of (a) precipitation and accretion through colloidal processes or (b) events in which organisms play a significant role?
- Which factors determine the shell structure of the nodules as well as the different chemical composition of the various zones in the nodules? What clues regarding their formation can we derive from nodules that exhibit clear differences between the upper side exposed to the seawater and the lower side exposed to the sediment with an equatorial bulge located between the two?
- What is the significance of the nucleus of the nodules? Do the encrusted fragments of rock provide us with information on the geological events connected with the nodule formation and the supply of elements?

We may draw the following conclusions from the foregoing discussion (Friedrich et al., 1977):

1. The increased heavy metals in seawater close to the seafloor as well as the high proportion of heavy metals bound to colloids indicates that the heavy metals are precipitated from the seawater with the particulate of the colloids.

2. The preferential occurrence of heavy metals in the hydrogenous phase of the deep-sea sediments and their enrichment in the fractions <100 μm indicate con-tinued growth of the colloidal-disperse phase in the deep-sea sediment with the formation of Fe-Man concretions in the μm range.

3. The occurrence of countless micronodules with clearly detectable zonal structure and differences in the distribution of elements from zone to one indicates a further phase of nodule growth. Microorganisms may participate in the formation of the nodules.

4. The macronodules, which are frequently characterized by typical upper and lower sides as well as by an equatorial bulge, form preferentially at the seawater/sediment interface with Fe–Mn hydroxides building up around micronodules or around foreign components acting as concentration nuclei.

5. The occurrence of rock fragments with an alteration zone as nodule nuclei (Roonwal and Friedrich, 1975) and the large number of seamounts composed of basaltic-andesitic material, indicate that the metal was also in part supplied by submarine volcanic activity. Metallic spherules with high Ni contents (Friedrich and Schmitz-Wiechowski, 1975), which are presumably cosmic in origin, are also possible sources of metal, in particular for Ni. It was observed that the metal phase of these spherules was displaced by hydroxides while the magnetite outer shell was retained.

6. Mechanical and chemical changes on the lower side of the nodules indicate that the macronodules may undergo destruction. The low Fe content in the outer zones of the nodules, which are in contact with the seafloor and the low Fe content in the material abraded from the nodules, is a particularly striking feature. A uranium deficiency was also determined in these zones.

Further detailed geochemical studies of macro- and micro-nodules from areas with different sediment-facies characteristics are needed in order to permit definitive statements about the formation of the nodules and the mining of the fields of nodules.

2.7 Mining Claims and Exploration

After 1987, on the basis of The United Nations Convention on the Law of the Sea, France, Russia, India, Japan, Inter-ocean metal organizations (Bulgaria, Cuba, Czech Republic, Poland, Russian Federation, and Slovakia), People's Republic of China, and the Republic of Korea applied to the United Nations and acquired their own mining claims. India acquired a claim in the Indian Ocean, while the other nations acquired their claims in the C–C zone in the Pacific Ocean. On November 16, 1994, the United Nations Convention on the Law of the Sea came into effect and at the same time, the International Seabed Authority was started in Jamaica to supervise the deep sea and its resources. The authority is now preparing mining codes and guidelines for marine environmental surveys, which has become an important issue. In any case, commercial mining may only be carried out after careful consideration of the effects on the environment and with international cooperation.

The objective for the exploration of deep-sea mineral resources is to confirm the amount, grade, and abundance of such nodules. A high-precision bathymetric chart must be prepared to search and fix the location of manganese nodules on a bathymetric chart. It is the most important part of the exploration. Therefore, a research vessel is installed with high-precision positioning system and echo sounders. Systems such as Sea Beam of the United States and Hydro-sweep of Germany are aboard such vessels.

Water pressure on the deep seabed is very high. For example, at 5000 m water depth, the pressure may be as high as 50 atm, with a water temperature as low as 1 to 20°C. It is an extreme environment. Since almost no light, nor electric waves, can be used, equipment that can be operated by ultrasonic waves is most effective. Therefore, research vessels are installed with ultrawave systems for research activities.

In order to conduct exploration, highly precise data and information on the horizontal and vertical variation of grade and distribution of manganese nodules must be effectively obtained. Of course, exploration must be carried out using different methods and equipment depending upon the characteristics of the target ore types. In order to determine the origin of resources, much data and information are needed. For this purpose, it is important to collect basic data through keeping a gravity meter and a proton gravity meter in operation all the time during the navigation and exploration.

To explain the exploration methods of manganese nodules, the survey system of the Metal Mining Agency of Japan is introduced here as an example. The agency has been exploring for manganese nodules for more than 20 years in the Clarion–Clippertone Zone southeast of Hawaii.

In order to draw bathymetric charts of vast sea areas effectively, they have installed the hydro-sweep (multibeam echo sounder) made by Kruppe Atlas Co., Germany. Using this system, while sailing at 8 to 10 knots, they can acquire geographical information of the seabed, which is twice as wide as the depth of the water and can simultaneously draw bathymetric charts by computer on board the vessel.

For automatic measurement of the abundance and diameters of manganese nodules, ultrasonic waves from a narrow beam echo sounder (30 kHz), a precision depth recorder

(12 kHz) and a narrow beam subbottom profiler (3.5 kHz) are simultaneously transmitted and the reflection of these three sound waves are analyzed by computer to estimate the abundance and the size of manganese nodules (Multi Frequency Exploration System).

To investigate the grade and continuity, manganese nodules and mud are sampled from the seabed with a freefall grab attached to a one-shot camera, a spade corer or a piston corer, and are then chemically analyzed. To understand the characteristics of the manganese nodules, observations of the seabed with a deep-sea TV and/or a steel camera are very important.

Navigation System	Exploration Equipment	Sampling Equipment
• Satellite Navigation	• Narrow Beam Sounder	• Free Fall Grab
• System Differential Global Navigation System	• Sub-bottom Profiler	• Spade Corer
	• Precision Depth Recorder	• Finder installed Power Grab
	• Multibeam Echo Sounder	• Drage Bucket
• Global Navigation System	• Multifrequency Exploration System	• Piston Corer
	• Finder installed Deep-sea Camera	• Boring Machine System
• Transponder System	• Side Scan Sonar	
	• Continuous Deep-sea Camera	
	• Conductivity-Temperature-Depth System	
	• Proton Gradio Meter	

During a comprehensive analysis of these research results, promising distribution areas are then selected for further exploration. In the selected areas, further research will be intensively carried out. In other words, sampling would be carried out at a narrower pitch to study more precisely the continuity and grade of manganese nodules. This brief outline is the basic flow of activities for manganese nodule exploration.

2.8 Indian Ocean Nodules

Chemical analyses of manganese nodules obtained from the Indian Ocean floor have enabled the demarcation of Indian Ocean into regions containing nodules of distinctive composition. It must also be mentioned that as far as the Indian Ocean is concerned, the amount of nodule samples hitherto available from various locations is still limited. Further, the bathymetric of the Central Indian Ocean Basin, its nodules show smooth regional metal variation in them, although they vary from basin to basin (Table 2.1). From mineralogical point of view, the nodules comprise todorokite occurs principally in nodules that occur in basin areas, whereas delta MnO_2 (7 Å manganite, =birnessite) occurs principally in ferromanganese encrustation and is more common in elevated areas. Deposits from the elevated areas such as the Central Indian Ocean Ridge contain above average concentration of Fe, whereas those form the basins are rich in Mn, Ni, Cu, and Zn. Some nodules in basins are principally composed of birnessite. These observations reflect the redox condition of the depositional environment in which they accumulate and the nature of the metal supply, which determine the mineral forming reactions taking place. The mineralogical composition of the deposits exerts a strong influence on their chemical composition, delta MnO_2 (birnessite)-rich deposits being high in Fe and todorokite-rich deposits in Mn, Ni, Cu, and Zn (Table 2.3 and Figure 2.3).

TABLE 2.3

Average Metal Concentration on Weight % of Nodules from the Indian Ocean

% Element	Central Indian Basin	Wharton Basin	South Australian Basin	Seychelles Basin	Agulhas Basin
Mn	26.1	19.8	22.6	25.0	15.9
Fe	7.6	11.2	10.9	16.5	14.1
Ni	1.20	0.65	0.96	0.58	0.66
Cu	1.16	0.54	0.49	0.18	0.15
Co	0.12	0.21	0.15	0.36	0.31
Ni + Cu + Co%	2.48	1.40	1.60	1.12	1.12

Source: From Roonwal, G. S., 1981. *Manganese nodules in the Indian Ocean.* Sci. Rep., New Delhi, 18: 384–391.

The causes of nodule variability in the Indian Ocean, particularly between different basins are complex. They depend upon the nature of source of the metals, be it from normal sea water, digenetic recycling of metals through the interstitial water of the sediments, or submarine volcanic activity. In addition, the depositional environments play an important role in determining nodule variability. Not only do the redox potential and nature and amount of sedimentation influence the mineralogy of the deposits, but also factors such as the depth of calcium carbonate compensation (CCD) and bottom water characteristics are of importance in determining their composition. Near to or at the CCD, there is a potential supply of elements to the deposits via the dissolution of sinking organic remains, which may partly explain high metal values in nodules from basin areas. Bottom water characteristics and circulation patterns are not fully known throughout the Indian Ocean, but it is believed that the Antarctic Bottom Water (AABW) influences both nodule abundance and composition, as in the Pacific Ocean (Figure 2.4).

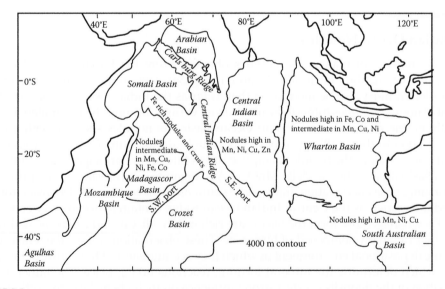

FIGURE 2.3
Shows the regional metal contents of the manganese nodule in the Indian Ocean. (After Roonwal G. S., 1986. *The Indian Ocean: Exploitable Mineral and Petroleum Resources.* Springer Verlag, Heidelberg, XVI, pp. 198.)

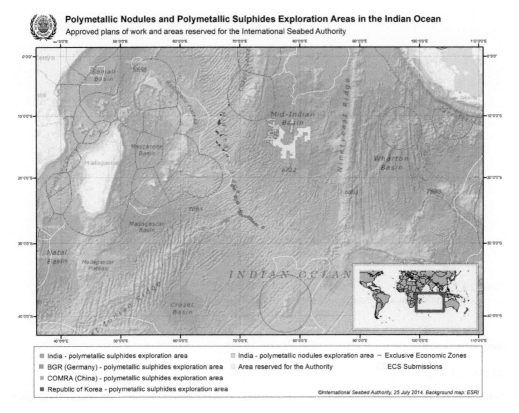

FIGURE 2.4
Indian Ocean showing claims for nodules and sulfide exploration.

In the Indian Ocean, promising economic grade occurs in the following basins, which have been arranged according to decreasing values of (Cu + Ni) content. (i) The Central Indian Basin (W of 90° east ridge), (ii) The South Australian Basin, (iii) The Wharton Basin (Keeling Basin, E of 90° east ridge), (iv) The Seychelles Basin, (v) The Agulhas Plateau Basin, and (vi) The Mozambique Basin (Roonwal, 1981).

Thus, it emerges that within the Central Indian Basin also there is difference in the composition of nodules west of 90°E ridge from those from east of the ridge (Table 5.1 and Figure 2.5).

Several reconnaissance cruises have taken place in the Indian Ocean. Some of these were as part of the International Indian Ocean Expedition (1954–1964), and some within the scope of Deep Sea Drilling Project (DSDP). Nevertheless, data regarding Indian Ocean nodules is scanty or unpublished. There is also equally scanty information on the bottom topography of the Indian Ocean floor. Near land, some nodules and encrustation have been reported also from the Carlsberg Ridge of the Arabian Sea. However, good nodules have been found only in the deep basins far from land and comprehensive review has been given by Frazer and Wilson (1979, 1980) of the Scrips Institution of Oceanography in the United States and a very detailed survey has been conducted by the Ministry of Earth Sciences (earlier Department of Ocean Development, Government of India), and the National Institute of Oceanography, based in Panjim, Goa. These surveys have identified mineable areas in the Central Indian Basin. India has since been given a "Pioneer Investor" status by the International Seabed Authority (ISA website).

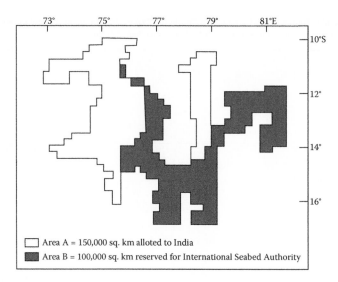

FIGURE 2.5
Area allocated to India as a Pioneer Investor in the Indian Ocean (www.isa.org.in).

References

Bonatti, E., 1982. Hydrothermal metal deposits from ocean rifts: II classification. In: P. A. Rona, K. Bostrom, L. Laubier, K. L. Smith Jr (eds.). *Hydrothermal Processes at Sea Floor Spreading Centres.* Plenum, New York, pp. 491–502.

Bonatti, E., Kraemer, T., and Rydell H., 1972. Classification and genesis of sub-marine iron-manganese deposits. In: D. R. Horn (ed.). *Ferromanganese Deposits on the Ocean Floor.* IDOE, Nat. Sci. Foundation, Washington, DC, pp. 149–902.

Frazer, J. Z. and Wilson, L. L., 1979. Manganese nodule deposits in the Indian Ocean. SIO Ref 79-18, *Scrips Inst Oceanogr,* pp. 71.

Frazer, J. Z. and Wilson, L. L., 1980. Manganese nodule deposits in the Indian Ocean. *Mar Min,* 2/3: 257–291.

Friedrich, G., Pluger, W., Kunzendorf, H., Roonwal, G. S., Schmitz-Wiechowski, A., Gurkran, A., Zuleta, J. R., and Kromer, E., 1977. Studies on the geochemistry and genetic interpretation of manganese nodule deposits. *Nat. Resour. Dev.,* 6: 26–47.

Friedrich, G. and Schmitz-Wiechowski, A., 1975. Mikroskopische und mikro-analytische Untersuchungen an metallischen kugelchen und mangan-mikroknollen aus tiefseesedimenten des pazifischen ozeans (micro-scopic and microanalytical studies of metallic spheres and manganese micronodules from deep sea sediments in the pacific) BMFT-For-schungsber. Meeresforschung M 75-02; pp. 8–31.

Glasby, G. P. (ed.). 1977. *Marine Manganese Deposits.* Elsevier, Amsterdam, p. 523.

Horn, D. R., Horn, B. M., and Delach, M. N., 1973. Copper and nickel content of ocan ferromanganese deposits and their relation to properties of the substrate. In: M. Morgenstein (ed.). *The Origin and the Distribution of Manganese Nodules in the Pacific and Prospects for Exploration.* Valdivia Manganese Exploration Group and Hawaii Institute of Geophysics, Office of the International Decade of Ocean Exploration, Washington, DC, pp. 77–83.

Mero, J. L., 1965. *The Mineral Resources of the Sea.* Elsevier, Amsterdam, pp. 312.

Murray, J. and Renard, A., 1891. *Deepsea Deposits Rep. Sci Results Explor Voyage HMS Challenger, 1873–1876.* HMSO, London, pp. 525.

Roonwal, G. S., 1981. Manganese nodules in the Indian Ocean. *Sci. Rep.*, New Delhi, 18: 384–391.

Roonwal G. S., 1986. *The Indian Ocean: Exploitable Mineral and Petroleum Resources.* Springer Verlag, Heidelberg, XVI, pp. 198.

Roonwal, G. S. and Friedrich, G. H., 1975. Studies on basaltic grains and phosphorite matrix in the core material of deep sea ferromanganese nodules from the Central Pacific; BMFT-Forschungsber. Meeresforschung M., 75-02; pp. 151–170.

Somayajulu, B. L. K., Heath, G. R., Moore, T. C., and Cronan, D. S., 1971. Rate of accumulation of manganese nodules and associated sediment from the equatorial Pacific. *Geochim. Cosmochim. Acta*, 34: 621–624.

Sorem, R. K., 1973. Manganese nodules as indicators of long-term variations in sea floor environment. In: Morgenstein (ed.). *The Origin and the Distribution of Manganese Nodules in the Pacific and Prospects for Exploration.* Valdivia Manganese Exploration Group and Hawaii Institute of Geophysics, Honolulu, Hawaii, pp. 151–164.

Sorem, R. K. and Foster, A. R., 1972. Internal structure of manganese nodules and implications in beneficiation. In: D. R. Horn (ed.). *Ferromanganese Deposits on the Ocean Floor.* IDOE, Nat. Sci. Foundation, Washington, DC, pp. 167–179.

3

Ores in the Deep Sea: Cobalt- and Platinum-Rich Ferromanganese Crusts

3.1 The Cobalt-Rich Crusts

Ferromanganese encrustations or crusts have been known together with manganese nodules for their Ni–Cu–Co metal content. Recent researches have indicated that these crusts occurring on ancient volcanoes have metal enrichment different from those found in the nodules. They show high levels of cobalt (~1.5%) and enrichment of platinum (~1 g/ton). In addition, a group of other elements, such as rhodium, cerium, and titanium, which are of strategic significance occur in them. The earliest ferromanganese was dredged for the first time during the famous HMS *Challenger* Expedition (1872–1882). On board RV *Sonne*, a team of scientists recovered such crust from the Mid-Pacific Mountain (MPM) and Line Islands in 1981 and 1984 (Halbach 1982, 1989). During 1983 and 1984, on board RV *S. P. Lee* of the United States Geological Survey (USGS) investigated crusts from the Nector Ridge (near Hawaii Islands). The University of Hawaii also undertook cruises around the Hawaiian Islands well within EEZ limits of the United States. We know that the seafloor nodules are enriched in Cu, Ni, and Co in that order. But, in the ferromanganese crusts, cobalt is three to six times more abundant than that found in the nodules (Figures 3.1 and 3.2) (Manheim, 1986).

3.2 Mineralogy and Geochemistry

The cobalt-rich crust comprises mainly of Fe–Mn hydroxide. The dominant mineral is MnO_2 (Vernadite) with x-ray reflections at 24° and 1.42 Å. Carbonate fluorapatite has been found to be abundant in the inner part of several crusts. Amorphous $FeOOH \cdot xH_2O$ is the other dominant phase present along with δMnO_2 in all the crusts. In addition, the other common minerals include plagioclase, quartz, goethite, and minor minerals include barite, potash feldspars, calcite, zeolites, and clay minerals.

Delta-MnO_2 is the most oxidized phase of the three common deep sea manganese minerals, todorokite, birnesite, and delta-MnO_2. The source of this manganese is in the water column where manganese oxide occurs mixed with iron compounds in suspended colloidal flocs. The delta-MnO_2/$FeOOOH \cdot xH_2O$ composition of the crusts reflects a ratio in the water column. Geothite probably crystallized from the amorphous $FeOOH \cdot xH_2O$ phase in some places.

It is concluded that quartz and some of the plagioclases occurring in the crust appear to be of eolian in origin. The rest of the plagioclase as well as pyroxene, potash feldspar,

FIGURE 3.1
Cobalt-rich ferromanganese crust on an altered host rock. (Courtesy of Dr. Sanjay Kali, IMMT, Bhubaneshwar.)

zeolites, and others are considered to have been derived from the substrate or bed rock. It is possible to have been produced by resuspension of the weathered material. Calcite and phosphorite are authignic minerals and suggested to have been derived from biological activity in the surface water. The amount of phosphate present could reflect the intensity of upwelling around the seamount (Hein et al., 1985, 1987a,b).

The chemical composition of individual samples is significantly different, but the average abundances of these metals are remarkably similar for large areas, at least in the central Pacific, where only such encrustation have been investigated. Data from the Pacific Ocean (Table 3.1) show a range mostly between 20% and 30% Mn; 15% and 20% Fe; 0.5% and 1% Co; 0.30% and 0.70% Ni; 0.10% and 0.25% Pb; 1.0% and 1.5% Ti, and 0.50% and 0.15% Cu. In the border land and marginal basin areas, these metals are diluted with a significant contribution of detrital minerals, which increases the Si and Al contents. According to the data available, it could be found that the Fe/Mn ratio is greatest for island areas and border land areas, which may partially explain the lower amount of Co, Ni, and Pb, which are associated with Mn. The Ti values are also low for these areas.

Examination of the crusts has shown that Co is commonly enriched in the outer layers of crusts, whereas Pt is enriched in the inner layers. Co, Pt, Ni, Ce, As, Mo, Cd, and Zn concentrations are possibly correlated with Mn content, and Cu Ce are possibly correlated with Fe. Al, Si, Ti, and K comprise the alumino-silicate fraction, with a part of Ti as well as V, Sr, and Pb partitional between the iron and manganese oxide phases.

3.3 Processes of Accretion and Formation

From the available information, it is apparent that cobalt-rich crusts are hydrogenetic. All the metals, except minor contributions from detrital phases, are derived from sea water. It appears that substrates do not contribute to crust chemistry due to the lack of correspondence between substrate type and the crust composition. Likewise, hydrothermal

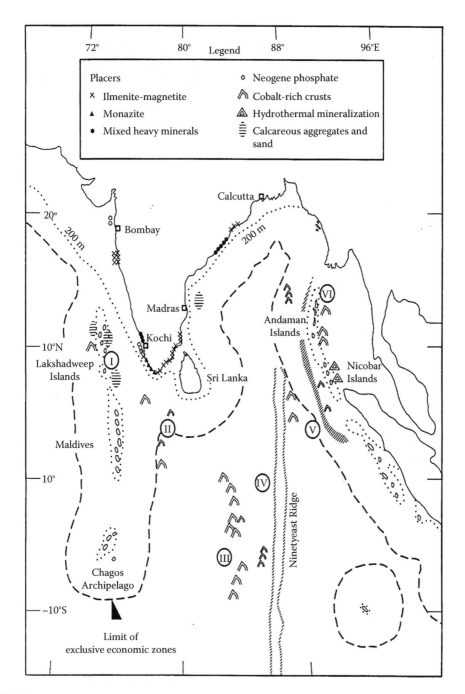

FIGURE 3.2
Major mineral occurrences in the northern part of the Indian Ocean. (After Roonwal, G. S., 1997. *Mar. Georesourc. Geotechnol.* 15: 21–32.)

TABLE 3.1

Average Chemical Composition of Various Elements in Crusts for Pacific Ocean Areas

Areas	Mn	Fe	Co	Ni	Cu	Pb	Ti	Fe/Mn
Hawaii and Midway (Axis)	24	16	0.91	0.45	0.05	—	1.1	0.73
Johnston Islands	22	17	0.70	0.43	0.11	0.17	1.3	0.81
Marshall Island	21	13	0.74	0.45	0.08	0.14	0.9	0.61
French	23	12	1.2	0.60	0.11	0.26	1.0	0.56
Tonga Ridge	16	20	0.33	0.22	0.05	0.16	1.0	1.26
Average Pacific	22	15	0.63	0.44	0.08	0.16	0.98	0.81

Source: Hein J. R., Schwals W. C., and Davis, A. S., 1987. *Mar. Geol.*, 78: 255–283.

contribution could be ruled out because of the abundant minor metals (Co, Ni, etc.) present and a lack of hydrothermal manganese phases such as todorokite and birnesite in the crust is not evident.

However, what triggers the precipitation of the first molecular layer from sea water on to volcanic or substrate is still not fully understood. It appears that when the process of precipitation begins, it becomes perhaps autocatalytic. Suggestions have been made that iron was deposited first either on organism or encrusting on bed rock. It is possible that some other mechanism, such as iron, catalyzed the deposition of Mn. Hein et al. (1987) stressed the possibility that Mn^{3+} oxides may form, initially at the crust surface, followed by conversion to MnO_2. However, the relative amount of the oxide phases in crust is determined by their ratio in colloidal flocs suspended in the surrounding sea water. The crust chemistry of major element thus reflects both the particulate and dissolved sea water chemistry. In the water column, manganese and iron oxides probably occur together in the colloidal flocs. Within the oxygen-minimum zone, which ranges from 200 to 2000 m in depth, manganese is highly soluble and therefore also occurs in the dissolved form. The Fe/Mn ratio is smaller in the oxygen-minimum zone, which is below that zone. Thus, the manganophile elements Co, Ni, Pb would be more concentrated within crusts and flocs from the oxygen-minimum zone. It appears that the ultimate controls on the concentration of minor metals (heavy metals) in the crusts depend on (a) the concentration of each metal in the sea water, (b) amount of scavenging by the iron and manganese oxide phases, (c) mineralogy, which in turn depends on the Eh (redox potential) of the environment, and (d) dilution by detrital phases has been proposed by Goldberg (1954) and Bonatti et al. (1972). A slow, growth rate is conductive to a greater degree of scavenging by the major oxides.

The processes of autocatalytic and trace metals and ferromanganese compounds precipitate onto the crust surface, releasing the incorporated iron to and from the water column. The slower the growth rate, the more trace metals that would be fixed. Thus, the crust enriched in trace metals are those that have grown at the slowest rate. The flux of cobalt at all depth in the water column is apparently constant and the primary control on the degree of cobalt enrichment is the rate of growth. Hydrogenetic crusts are among the slowest formed deposits in the geological environment; the less rich cobalt contents in the outer part of the crusts indicate recent most growth rates were slower than the older and earlier growth rates. The decrease in cobalt and depth of water can also be attributed to rapid rates of growth due to additional amounts of iron incorporated into the crust. It is suggested that an increased supply of iron has been related to the dissolution of iron-bearing biogenic carbonate below the carbonate compensation depth (CCD). Whatever be the source of additional iron in the deeper water crusts, it does act as a diluting factor to the minor metals.

3.4 Crust Texture: Thickness, Rates of Growth, and Substrate Characteristics

Cobalt-rich ferromanganese crusts are found generally laminated and show a textural change that distinguish (a) an inner older generation of growth and (b) an outer younger generation of growth. The textural change is commonly marked by a very thin layer of phosphorite, which perhaps marks the depositional hiatus of ferromanganese compounds. The inner generation therefore represents higher density compared to outer layers. Further, it is observed that the inner layer may be impregnated by phosphorite, which may also contain laminae of detrital minerals or even large pieces of substrate rock. A botryoidal surface texture is an expression of the undulatory and columnar growth patterns of the crust. Individual laminae may be truncated and are distinguished from adjacent layers on the basis of more or less iron relative to manganese. In some places, botryoidal surface textures dominate, and at other places, crust surfaces are smooth. Botryoids could be of various sizes and may appear pronounced or subdued. In some cases, surfaces are polished and flatted from working by bottom currents or gravity currents carrying suspended sediment. Dark granular material occurs on undersides, sides or in the low areas of the botryoidal surfaces; this granular ferromanganese oxide is richer in cobalt than in the surrounding oxide.

Perhaps the thickest hydrogenetic crust reported in the literature is 24 cm. This crust dredged from 4830 m water depth appears to have grown during the past 65 my. The age of the substrate upon which it has grown, [10]Be dating has put the outer 16 mm growth at a rate of 2.7 mm/my, whereas the internal 16–38 mm grew at 4.8 mm/my; the age of the crust at 38 mm depth is 11 my. Most crusts, however, have maximum thickness of 8–10 cm and average thickness of crust is controlled by the length of growth and rate of growth, which in turn must depend on the oceanographic and geochemical conditions at the site of formation. Although only a few age data are available, it is getting clear that crusts are not as old as the substrates on which they grow.

The crusts should begin to grow without interruption once volcanism on the seamount ceases; growth continues unless crusts enter a reducing environment, either by burial by sediments or changes in the chemistry of the surrounding water. The oldest part of most crusts, may be 10–20 my younger than the edifices upon which they rest. Perhaps one reason for this age gap is that crusts grow in an unstable environment of the slopes of seamounts and ridges where gravity flow and mass movement destroy crusts. Crusts and substrates may, in fact, show two or three episodes of reworking, erosion, and regrowth.

It is thus also expected that the mineralogy of substrate would differ with rock type. Volcanic rocks, including breccia and volcano-clastic sandstone, contain plagioclase and pyroxene with minor olivine, magnetite, and the alteration products, phillipsite, clinoptilolite, hematite, goethite, maghemite, and calcite. Smectite and phillipsite are the dominant minerals in highly altered rocks. Phosphorite consists of carbonate fluorapatite and calcite with traces of quartz fine-grained sedimentary rocks, which mostly show smectite, zeolite, and plagiolase. Other minerals that occur less commonly are potash feldspar, illite, and chlorite minerals (Halbach et al., 1982; Hein et al., 1985). Basalt substrate is usually rich in

MgO and CaO and to some extent in Na_2O and Fe_2O_3. They are also found to be depleted in K_2O relative to volcanic breccia, reflecting the grain and loss of them during alteration processes, when phosphatic rock has been found to show high CaO and P_2O_5 contents, depending upon the degree of phosphatization.

More growth rate data are available for abyssal nodules than for hydrogenetic crusts. The outer layers or the younger generation of crusts commonly formed at 1–3 mm/my in contrast with nodules (1–10 mm/my) and hydrothermal crusts (1000–2000 mm/my). The inner crust layers or older generation may have grown at rates upto 5 mm/my.

3.5 Indian Ocean Occurrences

According to Roonwal (1988), several areas in the Indian Ocean have geological environment where exploration strategy could be employed. There are (a) old seamounts ($\geq$20 ma), (b) well-developed oxygen minimum zones, (c) areas of strong current, (d) ideally between 500 and 1500 m water depth, and (e) in equatorial zones (15–20°) latitude each side of the equator has been suggested and can be seen in Figure 5.1. In such exploration campaigns, one has to (a) avoid areas with atolls and coral reefs, (b) areas near continent, and (c) select areas with flat top of the seamounts for better picking/sampling and perhaps eventual detailed exploration to mining prospects, (d) the average crust thickness of more than 4 cm, cobalt grade of more than 0.8%, subdivided microphotography areas with extensive development of crust, (e) seismic (air gun) survey apply criteria to eliminate areas of slump, talus deposits, and sediment cover. Therefore, in the Indian Ocean, areas in proximity to India are worth considering (1) Lakshadweep Islands (only seamounts without reefs), (2) southern tip of Kerala, (3) seamount between 84° and 88°E at 10°S, (4) seamount at 86°E 14°S, and (5) SE Andaman around 91°E7°N and 93°–96°E and 10°–13°N. The areas can be extended to south of Ninetyeast Ridge, plus several locations covers, which may be within the EEZ of India (Roonwal, 1988, 1997, 2006).

References

Bonatti E., Kraemer, T., and Rydell, H., 1972. Classification and genesis of sub-marine iron-manganese deposits. In: D. R. Horn (ed.). *Ferromanganese Deposits on the Ocean Floor*. IDOE, Nat. Sci. Foundation, Washington, DC, pp. 149–902.

Goldberg, E. D., 1954. Marine geochemistry I. Chemical scavengers of the sea. *J. Geol.*, 62: 249–265.

Halbach, P., Manheim, F. T., and Otten, P., 1982. Co-rich ferromanganese deposits in the marginal seamount regions of the Central Pacific basin—Results of the MIDPAC-81. *Erzmetall*, 35(9): 447–453.

Halbach, P., Sattier, C. D., Teichmann, F., and Wahsner, M., 1989. Cobalt rich and platinum bearing manganese crust deposits on seamounts: Nature, formation and metal potential. *Mar. Mining*, 8: 23–40.

Hein, J. R., Manheim, F. T., Schwals, W. C., and Davis, A. S., 1985. Ferromanganese crusts from Noeker Ridge, Horizon Guyot and S.P. Lee Guyot: Geological consideration. *Marine Geol.*, 69: 25–54.

Hein, J. R., Morgenson, L. A., Clague, D. A., and Koski, R. A., 1987a. Cobalt rich ferromanganese crusts from the exclusive economic zone of the United States and nodules from the Pacific. In: *Geology and Resource Potential of the Continental Margin of Western North America and Adjacent Ocean Basin. Circum-Pacific Council for Energy and Mineral Resources.* Earth Science Series 6, Houston, pp. 753–771.

Hein, J. R., Schwals, W. C., and Davis, A. S., 1987b. Cobalt and platinum rich ferromanganese crusts and associate substrate rocks from the Marshall Island. *Mar. Geol.*, 78: 255–283.

Manheim, F. T., 1986. Marine cobalt resources. *Science* 232: 600–608.

Roonwal, G. S., 1988. Cobalt and platinum-rich ferromanganese crusts in the Indian Ocean. *J. Geol. Soc. India, Bangalore*, 29: 358–361.

Roonwal, G. S., 1997. Marine mineral potential in India's exclusive economic zone: Some issues before exploitation. *Mar. Georesourc. Geotechnol.* 15: 21–32.

Roonwal, G. S., 2006. Cobalt and platinum-rich ferromanganese crust resources in the marine environment. *Indian J. Geochem.* 21(2): 495–506.

4

Ores in the Deep Sea: Seafloor Massive Sulfides and Metalliferous Mud

4.1 Introduction

The discovery of hot plumes and black smokers as observed by *Alvin* submersible in 1979 has been an exciting discovery. Observation of rapid flow of hydrothermal fluids of temperature as high as 350°C were recorded from the ocean floor. In the as tall as 13 m mineralized chimneys on the seafloor, clouds of affluent black smoke with suspended metal sulfide minerals puffed from the top of the chimneys. Equally exciting is that such black smokers or chimneys support different type of fauna, which has grown around them due to nutrient available. At present a large number of locations on the midoceanic ridges have been located in the eastern Pacific Ocean, from north to south, in the Fiji Basin, in the Lau Basin, in the Sea of Japan, Okinawa Trench, the Mid-Atlantic Ridge, and lastly and more recently the first discovery of massive sulfide in the Indian Ocean in 1994. In January 1994 the German ship RV *Sonne* recovered massive sulfide from the triple junction in an area called the "Sonne field."

Marine scientists are now busy investigating the formation of such hydrothermal plums in varying geological settings such as constructive plate boundaries (East Pacific Rise, Mid-Atlantic Ridge, Mid-Indian Ocean Ridge) or destructive plate boundaries (Fiji Basin, Lau Basin, Okinawa Trench). In the field of marine geology, geophysics, and chemistry, seafloor hot springs are revolutionizing models in the oceans. Marine biologists who study the unique biological communities around the hot springs are revolutionizing many conventional notions of the deep-sea animals.

More marine surveys and research conducted during the past few decades has led to the discovery of seafloor massive sulfide deposits (SMS). These SMS deposits are high-grade, localized, and form in areas of active undersea volcanism. They are found generally at a depth of 2000–3000 m. Such observation of SMS deposits has given the earth scientist an opportunity to study primary ore-forming hydrothermal systems in real time. Besides, this system supports a unique and varied fauna.

SMS have a wide range of minerals but mostly as sulfides. Their high metal values and grade make these an economic resource. This created interest in SMS an alternative source of metals, in addition to the known land-based resources. The available results to date have been encouraging. As more and higher-grade deposits are discovered, commercial interest in SMS is likely to grow in future. It is thus important to consider available technological systems or that are likely to be developed to carry out commercial mining. At present, several sophisticated systems have been developed with a view to extracting manganese nodules at depths of 2000–6000 m. The Red Sea project form mining of metal-rich mud in the deeps has proved the viability of pumping metalliferous mud from over 2000 m depth.

At this time, much emphasis is given to mining systems developed for SMS mining, as can be seen in the work of several consortia (websites). The Red Sea mud and manganese nodules, SMS, are hard consolidated deposits, similar in form to land-based sulfide ore bodies that are quite different from the Red Sea mud. Thus their mining method needs to include a device for disaggregating the deposit prior to transport to the surface. Some alternatives have been considered, such as solution mining or vent capping (Chapter 10). However, physical crushing of the ore is still considered as the first choice by most likely solution by engineers involved in the engineering solutions focussed to the problem. In addition to the technology, there are other aspects to consider such as world demand for base metals, the status of deep-sea deposits in international law, and environmental considerations.

4.2 Nature, Distribution, and Origin of SMS Deposits and Metalliferous Mud

Hydrothermal ore formation is controlled by processes at the spreading centers of the ocean. Each spreading segment is characterized by deep transform fault intersections caused by increase in cooling and resulting heat loss in the upwelling mantle material. The metallic ore formation system involves circulation of seawater through the fractured volcanic rock of the oceanic crust. The heating of the seawater is carried out during passage through newly emplaced hot plutonic and volcanic rocks. Depending on the degree of mixing of the hot aqueous metal-rich hydrothermal solution with normal seawater, chemical exchanges between seawater and oceanic crust result in the transfer of certain transition metals (Cu, Fe, Mn, Ni, Zn, etc.) from crust to seawater and certain alkaline earth metals (Mg, K, B, Rb, Cs, etc.) from seawater to sulfides of Cu-Fe crust. Polymetallic sulfides of Cu, Fe, and Zn precipitate from the unmixed hydrothermal solutions, which are at high temperatures (350°C) and under acidic and reducing conditions when they come in contact with the cold seawater.

The important indicators are geological, petrological, geophysical (acoustic, seismic, electrical, and magnetic) and biological. Petrographic study of associated sediments indicates hydrothermal origin and compositional types in addition to their isotopic characters. Hydrothermal sediments contain high proportions of transition metals such as Fe, Mn, Cu, and Zn, low Al and high Si/Al ratios and Fe/Mn ratios, between 1 and 200. Base metal sulfides are associated with anomalous $^{234}U/^{238}U$ ratios, have isotopic compositions at equilibrium with seawater, and Pb isotopic compositions resemble those of oceanic ridge tholeiitic basalts. The rare earth element pattern is similar to seawater, and a higher U and lower Th content than in normal pelagic sediments characterize metalliferous sediments.

Certain geochemical processes can also reveal the presence of metalliferous deposits. Geochemically in the hydrothermal seawater interaction, low dilution favors direct discharge of hot, acid solutions enriched in metals as in the Galapagos spreading center. An excess of rare He stable isotope, ^3He, above oceanic background together with an excess of unstable ^{222}Rn isotope may indicate discharge areas of hydrothermal circulation. Methane and hydrogen concentrations in 21°N EPR fluids are greater by 10^5 times than normal deep-sea concentrations. Higher suspended particulate matter together with high silica saturation with respect to seawater background is a strong evidence of hydrothermal discharge. Moreover, the sum of Fe_2O_3 and MnO_2 components shows a distinct increase in the direction of the vent following the same pattern as the percent of carbonate-free matter. Hydrothermal mineral deposits are influenced by crystal structure, permeability, rate of spreading, and distribution of heat sources. Structurally they occur along block faulted topographic highs adjacent to rift valley, volcanic seamounts on oceanic ridges, fracture zone that transect spreading centers, and basins along axial trough present during the rift stage in opening of the ocean basin. The faults and associated fractures that enhance the permeability of the rocks facilitate hydrothermal circulation.

Midoceanic hydrothermal sulfide environment is associated with characteristic patterns of convective and conductive heat flow. The highest value of heat flow occurs at discharge zones and the convective heat flow is manifested according the water circulation. Under restricted circulation the temperature of near bottom water may increase by 10°C. However, under conditions of open circulation the thermal anomaly is insignificant and intermittent conductive heat flow in sediments near the axis of spreading ridge shows a large variability between adjacent values (0 to 30 heat flow unit) and high average flux (2.5 HFU). The presence of contrasting acoustic impediment between seawater and hydrothermal fluids will produce a reflecting interface for echo-sounder waves and this will reveal the presence of metallogeny. In a similar fashion, a hydrothermal field is identified by shallow focus microearthquakes along the rift zones. Even the earthquake, which swarms along spreading centers in Atlantic and Indian oceans, is associated with seafloor metallogeny. Self-potentials involving a gradient of redox potential are often encountered with metalliferous sediments. A distinct low value of residual magnetic intensity by alteration of the magnetic minerals in the basalts along a high-temperature discharge zone constitutes an important criteria in recognition of such sites. Hydrothermal systems providing melliferous minerals, sediments, and crust are located in three principal tectonics environment: (a) on the spreading ridges which encircle the globe, (b) along volcanic arc fracture and subduction zones related to them, and (c) on seamounts of volcanic center. Biologically, evidences at Galapagos spreading ridge and at the EPR 20°S reveal the presence of rich and diverse fauna of giant clams, mussels, crabs, sea anemones, limpets, and chitons whose presence is related to proliferation of bacteria in H_2S-rich solutions emanating from the vents. Si derived from hydrothermal discharge may also support varied population of silica discharge as in TAG field. As a consequence of hot fluids passing through the organic rich mud, occurrence of hydrothermal petroleum is reported from the vents, as in the Sea of Cortez. The materials condensing at the seafloor give the mineral deposits a petroliferous odor.

Plate tectonic theory suggests that the outer surface of the earth is divided into around 10 "plates," which are irregularly shaped sections of earth crust of thousands of kilometers in size. The crust itself may be oceanic or continental in composition. While the oceanic crust is predominantly basaltic, dense, and relatively thin (10–12 km), and forms a continuous skin around the earth, the oceanic crust, this oceanic layer in many places is overlaid by a less dense granitic layer. The average thickness forming continental crust is about 23–35 km. Further, we know that each plate may be composed of oceanic or continental crust. It is normally a combination of the two. The plates are rigid masses of solid rock, but "float" on the relatively mobile mantle beneath, considered partially molten. The places subjected to various lateral movements are thought to be the result of convection currents taking place within the mantle. From the available information, we can divide the plates into three categories as given in Figures 4.1 and 4.2. It indicates (a) transform or

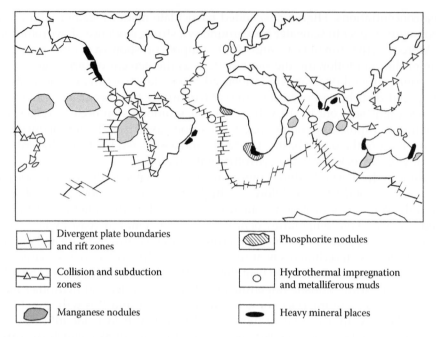

FIGURE 4.1
Major plate boundaries (compiled from various sources).

conservative boundaries are where one plate merely slips past another one, and crustal material is neither created nor destroyed, (b) convergent or destructive plate boundaries (subduction zones), which result due to oceanic crust being pushed into continental crust, resulting in the denser oceanic material being thrust into the mantle and destroyed, (c) divergent or constructive boundaries where new basaltic oceanic crust is created here. Two oceanic plates move apart and new basaltic magma (create new oceanic crust).

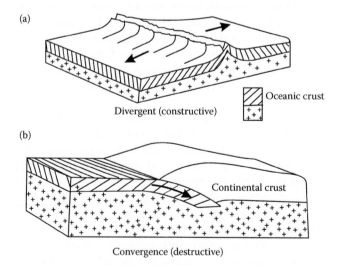

FIGURE 4.2
Showing concept of diverging and conversing plate boundary. (a) Divergent plate, and (b) conversion plate.

The diverging or spreading ridges areas of the ocean floor are characterized by high geothermal gradients (occurring large). Divergent plate boundaries represent the evolutionary sequence, and rifting may occur within a continental land mass, as we can see in the East African Rift Valley, or continental masses separate and seawater fills the rift, as in the case of the Red Sea today. The two land masses continue to move away from each other as new seafloor is created (Figure 4.10). This process may continue for several millions of years. Eventually the spreading rate slows down and may stop altogether. This mature stage is apparent on the present Mid-Atlantic Ridge and Carlsberg Ridge in the Indian Ocean. Spreading rates vary depending on the geological setting from 2 to 20 cm/year at the East Pacific Rise near the Eastern microplate, in the Indian Ocean—the mid-Indian Ocean Ridge is very slow-spreading ridge. It is at these spreading ridges that the vast majority of SMS deposits known today have formed. Figure 4.2 shows two schematic diagrams of such systems at oceanic ridges. Seawater enters the deep fissures present at the ridges due to flexing of the crust caused by vertical and lateral movements, and rapidly becomes heated by the high geothermal gradient. Seawater may descend to depths of up to 5 km and be heated to 300–600°C (Edmond, 1983). SMS deposits are the product of hydrothermal systems. As it percolates through the hot rocks, it leaches out metals are Mn, Fe, Co, Ni, Cu, An, Ag, Au, silicon, calcium, potassium, hydrogen, and sulfur. The superheated metal-laden brines will shoot back to the surface wherever possible. On reaching the seafloor, still at temperatures of up to 350°C, they emerge into a near-freezing temperature of 2°C abyssal sea water. This rapid cooling produces a drastic reduction in solubility and sulfide and oxide minerals precipitate out as a cloud of particles. It produces an effect known as "smoking." These vent emissions tend to be darker in color as the temperature rises, leading to a subdivision between "white smokers" (relatively low temperature), and "black smokers" (high temperature).

The minerals thus precipitated and produced as "chimneys" around the emitting vents such as spire-like chimney structures could grow to 20 or 25 m high (see Figures 4.9 and 4.7). Due to the friable, inhomogeneous nature of the sulfides, these chimneys are inherently unstable, further weakened by the effects of oxidation. Consequently they frequently topple over, leading to an accumulation of deposits composed of chimney fragments called the "basal mound." (Haymon and Kastner, 1981). Besides, the mineralization is widespread below the seafloor, in the form of a network of mineralized veins called a stockwork, though, as will be shown later, information on the vertical dimension of the deposits is extremely scarce.

A few important parameters are necessary for the formation of a hydrothermal ore deposit are (i) a source of metals, (ii) permeable pathways through the source rock, (iii) heat, (iv) water, and (v) a method of triggering precipitation to form an enriched deposit. The midoceanic diverging or spreading centers provide these factors. The basalt of the seafloor provides a rich source rock for transition metals, as mentioned previously. It is invariably fissured around oceanic ridges, providing permeable pathways. Heat is provided by the high geothermal gradient and water by the overlying 2000–3000 m of seawater. The precipitation of metals is initiated by the temperature difference between the exiting fluids (~350°C) and the ambient seafloor temperature (2°C). This leads to the necessary condition for ingredients of a hydrothermal system to occur along the whole length of the oceanic ridge system.

4.2.1 The Red Sea

The Red Sea is a spreading center at a very early stage of development (in geological terms). The African and Arabian landmasses are slowly moving apart along a rift running the full

TABLE 4.1

Resources of the Atlantis II Deep

Deposit	Grade—Data Based on the Dry-Salt-Free Material	Tonnage×16
Zinc	3.41 kg/m³ (2.06%)	1890
Copper	0.77 kg/m³ (0.46%)	425
Silver	6.7 kg/m³ (40.95 g/t)	3.75
Gold	0.512 g/t	0.047
Cobalt	58.53 g/t	5.368
Dry-salt-free material	–	91,700
Metalliferous sediments	–	696,330

length of the Red Sea. During the last 100,000 years, undersea volcanism combined with extremely saline pore water has leached metals from the surrounding bedrock and redeposited them as metal-rich mud. The metal rich mud collected in "deeps," depressions along the ridge axis where water temperatures may exceed 55°C due to localized volcanic activity. These deposits have been the subject of detailed study since 1966 (Degens and Ross, 1970), and at the time of writing, are closer than any other submarine sulfides to commercial exploitation.

The governments of Saudi Arabia and Sudan concluded a bilateral agreement, forming the Red Sea Commission (RSC) in 1974. The RSC was to share the costs and benefits of any mineral extraction operation in the Red Sea, including the mud. The commission subsequently retained the French Bureau de Recherches Geologiques et Minieres (BRGM) as geological consultants and the German company Preussag AG as technical contractors. Following detailed surveying and sampling, the Atlantis II Deep was selected as the site for a prepilot mining operation. The mud was sucked up from 2200 m depth using a vibrating suction head and submerged pumps. This initial operation concluded in 1981 was highly successful, and hopes are high that a five-year pilot operation using full-scale equipment will commence before 1990. Such a venture would produce an estimated revenue of $2240 million over the proposed 16-year life (see Degens and Ross, 1970 for details). The most important associated metals are copper, silver, gold, and cobalt. Drawing on the results of the prepilot phase Guennoc and Thisse (1982) estimated the total resources of the Atlantis II deposits to be as depicted in Table 4.1.

The Red Sea deposits may well be unique as no other site with comparable geological conditions is known to exist at present. The strong saline solutions associated with the evaporites, coupled with the limited oceanic circulation, has led to a rapid buildup of metalliferous sediments in a localized area. These sediments have the consistency of wet clay, unlike the deposits, which are hard consolidated sulfides more akin to cast iron.

4.3 Seafloor Sulfides in the Indian Ocean

The occurrence of active ridge metalliferous sediments in the Indian Ocean was first reported by Swedish oceanographer Bostrom and Peterson (1966). They studied samples from Mid-Indian Ocean Ridge and Marie Cleste Fracture Zone, which offsets the ridge north of its triple junctions. Fe–Mn encrustations were collected from the valley wall and sulfides were detected in the vesicles of the basalts. The associated Si and Ca clays also showed a strong influence of hydrothermal activity.

Cann et al. (1977) has reported hydrothermal Fe–Mn encrustations on the basaltic lava from near Gulf of Aden in the form of spongy, brown MnO coating over friable yellow to green smectite. Rozanova and Baturin (1971) have reported stockwork-type Cu–Fe sulfides in larger offset, ridge–ridge transform fault segments.

Preliminary reconnaissance of Carlsberg Ridge between Owen Fracture Zone (lat. 9°50′N) and Vityaz Fracture Zone (lat.5°17′S) and the Central Indian Ocean Ridge between 5°N and 20°S have observed possible hydrothermal activity. Around 1° 40′S-67° 46′E and 5° 21′S-68° 37′E in a structurally disturbed zone at the intersection of the west wall of the rift valley with the Vityaz Fracture Zone, mineralization in the form of disseminated grains and veins of Cu–Fe sulfides (Chalcopyrite) occurs in a matrix of basalt hydrothermally altered to greenschist facies. Higher levels of He^3 values in the Carlsberg Ridge zone of Indian Ocean correlate with a higher Ch_4 content. These occurrences have been reviewed by Roonwal and Mitra (1988). In 1993, first massive sulfides were collected from the Sonne hydrothermal field near the Rodriques Triple Junction (Halbach et al., 1994, 1996).

The Red Sea sulfides areas are an example of mineralization at an early stage of slow-spreading center of the ocean floor, within a closed basin with restricted circulation. They occur in a number of deeps in the Red Sea where transform faults intersect the Median Valley. When seawater comes in contact with the thick layers of rock salt, buried in the seafloor, it acquires high metal transporting capacity. The hot metal-rich hydrothermal solution, when ascending to the top, comes in contact with the cooler water and under reducing conditions precipitates the sulfides in a stratified form. Their deposition is controlled by transform faults.

The Red Sea sulfides occur in two facies: (a) mixed sulfides of Fe, Cu, and Zn and (b) pyrite. The former occurs as gray, fine-grained, massive beds, principally made of sphalerite together with pyrite, chalcopyrite, and marcasite. Bignell (1978) analyzed the Atlantis II Deep sediments and gave a value of Zn, Co, and Fe over 10%, 2%, and 15% respectively. The metal is supposed to have been derived by leaching from volcanic rock and transported by brine. The pyrite occurs within the Atlantis II Deep and contains a higher concentration of Fe up to 29% with Cu 1.3% and Zn 3.4% with minor amounts of Au and Ag. One may refer to Degens and Ross (1969) for a fuller account of them.

At a time of growing scarcity of critical minerals, the discovery of hydrothermal metalliferous sulfide deposits associated with divergent plate boundaries along the ocean floor is of much significance. These deposits in the oceanic crust hold significance in the context of mineral exploration. The long-term significance is the actual recovery of metals from these formations, which, however, involves development of new technology at high costs. Nevertheless, the sulfide deposits constitute an important source for future metal needs and the oceanic crust in which they occur constitute a large part of the ocean bottom. Their short-term significance lies in the fact that the findings on large slices of oceanic crust can provide practical guidelines for their commercial exploitation in the near future.

Seafloor sulfides were first discovered in 1979 on the East Pacific Rise. They were discovered after the UNCLOS. Therefore, UNCLOS primarily concentrated on "common heritage of mankind" for seafloor manganese modules are considered as a resource for copper, nickel, and cobalt. Though the metal content in them in low (approximately 2.5%), the huge quantity available on seafloor plain was considered enough for commercial mining. However, the discovery of seafloor sulfides on the midoceanic ridges is now considered more attractive for commercial mining on the seabed. The prime reason for this is their high metal content (Zn + Cu upto 30%) and concentrated occurrence, which would possibly make mining commercially attractive.

From the desire to understanding the earth to utilization of earth mineral resources for the benefit of society is a journey that continues to this day. Thus, earth scientists involved understand though Wegner's concept of "continental drift" to Hess's concept of "seafloor spreading" to the recent "plate tectonic" concept. Plate tectonics is a key important concept in understanding the earth history and metallurgy.

Broadly ocean floor comprises plains on which nodules and midoceanic ridges occur. The midoceanic ridges are the sites of occurrence of seabed sulfides. Again, there are two types of ridges and the ridges themselves define the plate boundaries. The first type is constructive margins, where new magma come out of the upper mantle and pushes the plates away. This is also called "diverging" plate boundaries. The second type is destructive plate boundary. Here two earth plates push each other: one goes under the other. This is called destructive plate boundaries and their zones are defined as "subduction." This is not a place to go into more detail of the process of both destructive and constructive plate boundaries. However, it is important to know that they occur in all oceans, the Pacific, the Atlantic, and the Indian.

In the Indian Ocean, constructive boundaries are on Mid-Indian Ocean Ridge, and Carlsberg Ridge, and the destructive boundary in the Andaman Sea. Therefore, both are important for locating seafloor sulfide deposits. The present synthesis concerns the main Indian Ocean. The Andaman Sea shall be discussed separately.

4.4 Volcanogenic Massive Sulfides (SMS)/SMS Deposits on the Mid-Indian Ocean Ridge

Earlier Indian Ocean ridges have been considered relatively less favorable sites for the exploration of hydrothermal mineralization because of their low to medium spreading rates. However, this view changed with the discovery of massive sulfides and vent biota in the TAG area of the Mid-Atlantic Ridge in 1985. Therefore, the GEMINO cruises by RV *Sonne*, a hydrothermal field was discovered on the Central Indian Ridge, about 200 km northwest of the Rodriquez Triple Junction (Herzig and Pluger 1988; Pluger et al., 1990). In the 1994 cruise in this area, massive sulfides were recovered (Halbach et al., 1994). Other indications of submarine hydrothermal activity in the Indian Ocean have been summarized by Roonwal (1986, 1997, 2005), Herzig and Pluger (1988), Roonwal and Mitra (1988), and Halbach et al. (1996, 1998). An overview of the tectonic pattern of the Central Indian Ridge is given in Figure 4.3.

4.4.1 SMS Deposits on the Mid-Indian Ocean Ridge

SMS deposits are defined as "massive" marine polymetallic sulfides meaning deposits that comprise more than 60% of sulfide material. The current theories of their mode of formation as discussed earlier have shown distribution of the largest known deposits, and finally, the details of the grades and likely global extent of the deposits. It is important to stress that the current database on SMS is no longer small. The vast majority to date have been located on active spreading centers, which extend for some 50,000 km across the floor of the world's oceans. However, around 1150 km of these systems have yet to be investigated in sufficient detail to reveal any SMS deposits that exist there. In addition, recent evidence suggests that back-arc basins and associated geological regimes may also host sulfide deposits. If this is so, it would greatly increase the global potential for SMS formation. In the India Ocean, India's Andaman Sea is seen as having this geological setting.

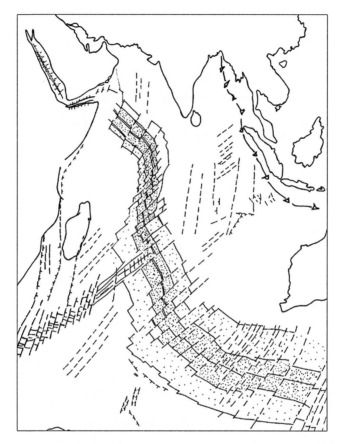

FIGURE 4.3
Ridge systems in the Indian Ocean (compiled from different sources).

Data on land-based deposits are thought to represent ancient SMS analogs. Occurring in the continent indicate that some form of stockwork mineralization may extend for 100 m or more concealed in the crust (Table 4.2).

The central Indian Ridge is the boundary between the African and Indian plates and forms a SSE-trending, midoceanic accretionary system in the equatorial Indian Ocean. To the north is the Carlsberg Ridge spreading center. Massive sulfide from the Indian Ocean (Halbach et al., 1998) show that a multiple hydrothermal event over a period of several ten thousands of years formed mineral occurrences, which are, more or less, north–south arranged in the MESO mineral zone of the Central Indian Ridge (Table 4.1). The site of this deposit is located in the central part of the fourth segment (about 270 km N of the RTJ) on a neovolcanic intrarift ridge. The particular stage of chemical and physical disintegration of the extinct sulfide deposit has been going on for at least 11,000 years. Therefore, the chimney structures have been quite destroyed by weathering processes. Three types of sulfide mineralization formed by hydrothermal venting processes have been distinguished by them as (1) Cu-rich massive sulfides (chalcopyrite-bornite-digenite-pyrite assemblage); (2) pyrite-marcasite massive sulfides; and (3) sphalerite-bearing jasper breccia. The main gangue phases are barite and amorphous silica.

On land, massive sulfide deposits of the Cyprus type (fossil midocean ridge massive sulfides) only rarely show a preservation of complete chimney structures; occasionally there is evidence of layered chimney fragments. Their studies results show that hydrothermal

TABLE 4.2

Typical Contents (wt.%) of Elected Metals from the Seabed: Sulfide Samples and Related Ancient Massive Sulfide Ores.

			Modern				Ancient	
			Juan de Fuca Ridge			Mid-Indian Ocean Ridge Cp Spl		
Location	21°N Active Vents	Cyamex Area	Northern	Southern	Axial Seamount		Cyprus	Canadian Precambrian
Zinc	32.3	40.8	6.3	54.0	19.2	0.14 / 31.10	0.2–0.5	5.0
Copper	0.8	0.6	0.5	0.2	0.13	31.6 / 0.71	2.5–4.0	2.0
Lead	0.3	0.05	0.1	0.3	0.4	–		0.0
Silver (ppm)	156	380	30	260	288	44 / 8	39	50
Gold (ppm)	0.17	–	0.08	0.13	–	890 / 260	0.3	181.76

Source: Adopted from Different Sources.
cp = Copper ore (Chalcopyrite).
sp = Zinc ore (Sphalerite).

chimneys are short-lived features that disintegrate after waning of hydrothermal activity. This is consistent with their observations that complete chimney structures are often lacking in the fossil environment of former midocean ridge hydrothermal sites. Moreover, assuming normal deep-sea conditions, chimney dissolution and disintegration have been place due to the disequilibrium of anhydrite and sulfide minerals with ambient oxidizing seawater. Only underlying massive and stockwork mineralization is preserved in the geological record (Table 4.3).

Mount Jourdanne on the South West Indian Ocean Ridge (SWIR) is another site where SMS have been discovered. On the Central Indian Ridge, one finds occurrence of SMS on the main CIR, Rodriguez Triple Junction (RTJ), on the eastern limb (SWIR) after Central India Ridge. Although, as mentioned above, serious efforts to locate hydrothermal activity on the Southern Central India Ridge (CIR) commenced in 1983 with the GEMINO program (Herzig and Pluger 1988; Pluger et al., 1990). However, again the main efforts were made in 1993 with the first discovery of the hydrothermal marine mineralization on the Cir (Halbach et al., 1998; Gamo et al., 2001; Jian et al., 2008; Kumagai et al., 2008). Banerjee and Ray (2003) and Wang et al. (2011) reported hydrothermal activity on the SEIR. The SWIR,

TABLE 4.3

Metal Values of Selected Sulfide from *Sonne* Field from Central Indian Ocean Ridge

			Chalcopyrite Type			Sphalerite-Jasper Type		
		Method	18–34	18–25	18–67	18–107	18–82	18–109
Zn	wt.%	ICP	0.07	0.04	0.14	31.10	23.40	27.30
Cu	wt.%	ICP	31.60	7.39	9.85	0.43	0.42	0.71
Ag	ppm	INA	33.00	44.00	11.00	<5.00	<5.00	8.00
Au	ppb	INA	220.00	890.00	240.00	250.00	240.00	360.00

Source: Halbach, P., et al. 1998. *Mineralium Deposita*, 33: 302–309.

TABLE 4.4

Licenses Overview for Seafloor Mineral Exploration in the Indian Ocean (ISA site)

Contractor	Draft Entry into the Contact	Expiry	Sponsoring State	General Location of the Exploration Area Under Contract	Type	Depth	Area (km²)
Govt. of India	March 25, 2002	March 24, 2017	India	Indian Ocean	Polymetallic modules	5000–5700	75000
China Ocean Mineral Resources and Development Association (COMRA)	Nov. 18, 2011	Nov. 17, 2026	China	South West Indian Ridge	Polymetallic sulfide	–	10000
Govt. of Republic of Korea	May 1, 2014	April 30, 2029	Korea	Central Indian Ocean Ridge	Polymetallic sulfide	–	10000
Govt. of India	Sept. 26, 2016	Sept. 25, 2031	India	Central Indian Ocean	Polymetallic	3000	10000
Federal institute of Geosciences and Natural Resources of Germany	May 2015	May 2030	Germany	Central Indian Ocean	Polymetallic sulfide	2600–3300	10000

is a very slow-spreading region, and was not considered attractive for locating hydrothermal mineralization. But the 1997 findings (German, 2003) gave evidence of hydrothermal activity here. Taking this into consideration, oceanograph expedition by Japan carried out a submersible survey in 1998 (INDOYO mission with Shinkai 6500 (Fujimota et al., 1999). An extinct hydrothermal mineralization was found at 2940 m in water depth at 27°51's/63°56'E) in the segment called Mount Jourdanne (Munch et al., 2000). An overview of exploration licences granted by the International Seabed Authority in the Indian Ocean for manganese nodules and seafloor sulfides is presented in Table 4.4.

4.5 Hydrothermal Mineralization in the Andaman Sea

One area within the EEZ that is potentially prospective for hydrothermal mineralization is the Andaman Sea. The Sumatra earthquake and the resultant tsunami havoc of December 26, 2004 have again focused attention on this region. Thus, it is important to review the mineral and hydrocarbon potentials of this region because, in the long term, it will be a site of large investments. The Oil and Natural Gas Corporation of India has committed nearly US$2 billion for the exploration of hydrocarbons in the Bay of Bengal. This is likely to increase in the coming years. Figure 4.4 shows concept of hydrothermal systems on Central Indian Ridge and Figure 4.5 shows generalize tectonics evolution of the Andaman Sea back-arc spreading centre.

The Andaman Sea in the northeast Indian Ocean lies between 6°–14°N and 91°–94°E (Figure 4.6a and b), is heavily sedimented (up to 6000 m), and has an actively extending back-arc basin on the Indo-China continental margin. In this respect, the Andaman Sea is an extensional basin, which began opening about 13 my ago (Middle Miocene), at a rate

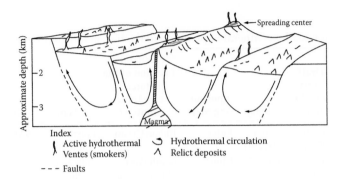

Index
⟨ Active hydrothermal   ∽ Hydrothermal circulation
⟩ Ventes (smokers)   ∧ Relict deposits
- - - Faults

FIGURE 4.4
Concept showing hydrothermal system on the Mid-Indian Ocean Ridge.

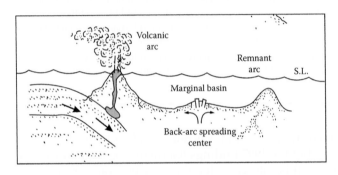

FIGURE 4.5
Generalized tectonic evolution of back-arc basin in the Andaman Sea.

of about 37 mm/yr (Curray et al., 1982). The total opening of the basin has been about 460 km and it now occupies about 800,000 km² (Rodolfo, 1969). Anomalous concentrations of Cu (up to 500 ppm), Zn (up to 500 ppm), and Pb (up to 900 ppm) have been reported in the coarse fraction (>800 mesh) of clay sediments taken at about 1500 m along the toe of a submarine valley, 12 km west of Narcondam Island (Banerjee et al., 1992).

The structure, tectonics, and geological history of the Andaman Sea are becoming better known (Biswas et al., 1992; Curray et al., 1982; Mukhopadhayay 1988; Rodolfo 1969). Magnetic anomaly data suggest that the Andaman Sea began to open about 13 Ma, with a spreading rate of about 37 mm/yr (Curray et al., 1982). Two islands (Barren and Narcondam) are volcanically active (Dutta 1991; Halder et al., 1992). The Andaman–Nicobar area can be considered a combination of a fan valley and an active spreading rift, which shows striking similarity to the Gulf of California (Curray et al., 1979), where sediment-hosted hydrothermal mineralization is well documented in the Guaymas Basin. By analogy, therefore, the Andaman Sea may be considered highly prospective for submarine hydrothermal minerals (Roonwal, 2009).

The bathymetry of the basin is complex. To the west lies the Andaman–Nicobar Ridge, which marks the boundary between the Indian and Eurasian-China plates, and to the east the Malay continental margin. The main topographic features of the basin are (a) the Central Andaman Rift, (b) the Central Andaman Trough, (c) the Deep Through, and (d) the Barren Seamount Complex, the Alcock Seamount, and the Sewell Seamount (Rodolfo, 1969). The deepest part of the basin is the central axial rift trending northeast–southwest and centered

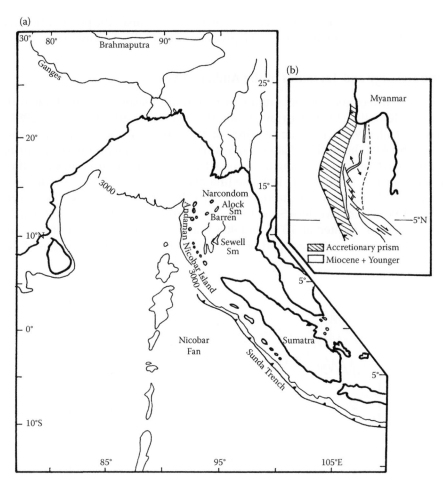

FIGURE 4.6
(a) and (b) shows location of Andaman Sea in the northeast Indian Ocean.

at 10.5°N: 94.5°E, at a depth of 4000 m (Curray et al., 1982). The rift valley can be traced to the south but is offset by NNW–SSE faults at places (Anon 1981) (cf. Curray et al., 1982). The main sources of sediment to the basin are the Irrawaddy, Sitting, and Salween rivers in Myanmar. The Irrawaddy presently discharges about 265×10^6 tonnes of silty clay per year, 90% of which is deposited in the continental shelf (Rodolfo, 1969). During periods of high level, the bulk of the sediment is trapped in the delta area or on the continental shelf. However, during low sea-level stands, they are introduced directly into the basin (Curray et al., 1982).

The matrix data show that there is no direct input of sediment from the Bengal fan into the Andaman Sea (Curray, 1991). Sediment in the Central Andaman trough is principally turbidity flows (silty clays), with an overall sedimentation rate of about 150 mm/1000 years (Rodolfo, 1969). The Central Andaman Rift is both an active spreading center and a fan valley analogous to that in the prospective for hydrothermal minerals (Roonwal and Glasby, 1996). The distal fan deposits have buried the rift valley in the north but have not penetrated to the southern part of the rift (Anon 1981; Curray 1991, 1982). The nature of neovolcanism in the spreading zone is unknown. Magnetic stripping in the southern portion indicates that salt is likely to predominate. However, the closest proximity of the spreading axis to the submarine active arc and the continental margin to the rest raises the possibilities that felsic volcanic may be present. Figure 4.6a and b gives location of the Andaman sea.

4.5.1 The Geology and Tectonics of the Andaman Sea

The geology and tectonics of the Andaman region are reasonably well known through the work of Rodolfo (1969), Curray et al. (1979, 1982), Mukhopadhyay (1984, 1988), and Biswas et al. (1992). Tectonic features of the Andaman Sea region are shown in Figures 4.7 and 4.8. The Indian Plate is underthrusting the long and narrow Burma plate (part of Eurasian China plate) along the Sunda Trench. The northern portion of the Sunda Trench is completely filled with clastic sediments of material from the Himalayan and Tibetan plateaus being brought through the north Indian river system. The Burma Plate carries the active volcanic western Sunda Arc. The site of the December 26, 2004 earthquake, whose epicenter was located at 3.3°N and 96.1°E and seabed to the west of it, is an accretionary prism that has been scrapped off the subducting plate, and is exposed, together

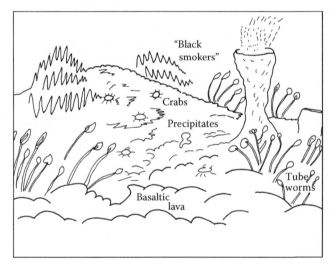

FIGURE 4.7
The formation of hydrothermal chimneys and precipitation of SMS.

FIGURE 4.8
Hand specimen of seafloor massive sulfide showing clearly mineralization and micro channels.

with Cretaceous ophiolites and various younger sedimentary units, on the uplifted islands of the Andaman–Nicobar Ridge. Major transform faults along the backbone of Sumatra (Sumatran fault system) to the south and cutting off the Eastern Highlands of Burma (Sagaing Fault) to the north are linked by short segments of back-arc spreading in the Andaman Sea. They constitute the boundary crust of the Andaman Sea, with the flanking volcanic arc and accretionary prism to the west and continental crust to the east being exposed on land in the Central Valley of Myanmar (Burma). This extensional zone was created by the north-westward rifting of the continental region represented by the eastern Burma highlands and the Malay Peninsula. A review of the land geology of the Andaman is given by Pandey et al. (1992).

4.5.2 The Nature of the Sediments

Data (Roonwal et al., 1997a,b) show that the sediment from the Andaman Sea consists mainly of quartz, chlorite, illite, smectite, kaolinite, and feldspar, with traces of calcite. These minerals occur in varying proportions. Sediments from the Central Andaman Trough contain higher amounts of illite than those from the flanks of the basin, which contain higher amounts of smectite.

The mineralogy of sediments from the eastern flanks of the Upper Nicobar Fan is more variable. Samples from a depth greater than about 2500 m consist mainly of quartz, illite, chlorite, smectite, and traces of feldspar, whereas those from shallower depths consist mainly of calcite, with the corresponding lower amounts of accessory minerals. Calculations of relative clay mineral abundances show that the Upper Nicobar Fan sediments contain about 30% smectite, 46% illite, and 25% kaolinite on average, whereas those from the Andaman Sea contain about 35% smectite, 39% illite, and 26% kaolinite, on average.

Data of the element concentration for the sediments from the Andaman Sea show them to be relatively uniform in composition, whereas sediment from the eastern flank of the Upper Nicobar Fan are much more variable in composition on account of the greater variability in the $CaCO_3$ content (3.87%).

4.5.3 Seafloor Spreading and Implication for Volcanogenic Massive Sulfides (SMS)

The discovery of "black smokers" since 1979 on midoceanic spreading ridges (Rona 1988) has initiated a major breakthrough in understanding how ancient massive base metal sulfides ore bodies were formed. However, neither the tectonic setting nor the basaltic nature of volcanism at midoceanic ridges are clearly relevant to the majority of the ancient massive sulfide ores, which tend to be associated with felsic volcanic rocks, apparently formed in the continental margin environments. At least three occurrences have been discovered on the modern seafloor, which are clearly analogous to the land deposit mentioned above. These are Jade deposits in the Okinawa Trough (Halbach et al., 1989), the Valu-Fa deposit in southern Lau Basin in the Back Arc Basin (Fouquet et al., 1991), and the PACHMANUS deposit in the Manus Basin in the Back Arc Basin near Papua New Guinea. The Andaman Sea is a substrate for back arc extension and is a continental rather than an island arc crust.

The main spreading axis of the Andaman Sea centers at 10.5°N and 94.5°E. This is a very well-defined linear, NE–SW trending, physiographic feature (Figure 4.9) at about 10.5°N. To the south, very short spreading and transform segments alternate; but on comparing magnetic lineation with present-day seismicity and topography (Curray et al., 1982), it was concluded that the multiplicity of spreading axes is becoming simplified into long continuous

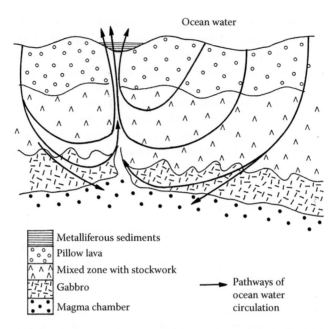

FIGURE 4.9
The initiation of hydrothermal system on the midocean ridges.

segments. To the north of the main spreading axis, the spreading zone becomes curved, and the spreading axis is oblique to the trend of the axis. Alcock Seamount, which forms the west margin of the axis here, is an uplifted block of young oceanic crust consisting of augite basalts (Rodolfo, 1969). Eguchi et al. (1979) concluded from tele-seismicity and gravity data that another spreading segment may exist centered at 14°N and 96°E (Figure 4.12), but is buried under a thick accumulation of clastic sediments.

The most prospective area for SMS is the main spreading axis, which is 5–8 km wide, and on the order of 500 m deeper than surrounding seafloor. At its northeast end, it is developed within very thick sediments (about 2300 m; Curray et al., 1979). Bathymetry (Figures 4.10 and 4.11) shows that the flat sediment surface at 3000 m water depth is transected by

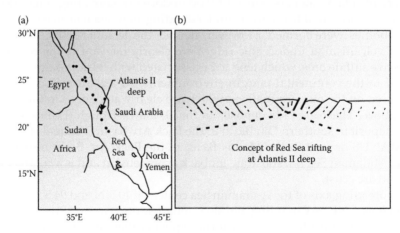

FIGURE 4.10
(a) Red Sea showing location of deeps and (b) concept of red sea rifting in the deeps.

FIGURE 4.11
Core sections of metalliferous sediments from Atlantis II Deep showing layers sulfides. (Courtesy: H. Backer, Germany.)

the rift valley. Upturned edges of sediments in seismic profiles suggested to Curray et al. (1979, 1982) that the spreading has been continuous, with episodic deposition of sediments coinciding with periods of Quaternary low stands of sea level. As a consequence, the NE part of the main spreading axis at about 94°2′E is heavily sedimented, whereas the remainder and the short spreading segments to the south are relatively sediment free.

Thus, in the Andaman Sea, there are possibilities of encountering hydrothermal venting in both a heavily sediment piled and a relatively sediment-free setting within the same spreading system. It may be noted here that the largest known seafloor deposits are in sedimented spreading centers. Why this is so, is not known; but it could be that the sediments act as a thermal insulation barrier to fluid flow, which results in a more focused discharge of hydrothermal fluid to the seafloor. The Andaman setting may provide answers to this question (Figure 4.12).

4.6 Industrial Interest in SMS

This section is based on the results of a postal and interview survey conducted to establish the views of private industry on the viability of MPS as an economic resource (Appendix I). Emphasis was placed on companies previously involved in examining the feasibility of manganese nodules extraction, as these clearly possess a distinct technological advantage over any would-be newcomers to the field of deep-ocean mining. The original consortia setup to develop nodule mining systems are shown in Tables 4.1 and 4.5. Representatives of each of these consortia were contacted and the overriding message was clear—further development of nodules has reached a hiatus and interest in SMS is restricted to a "watching brief," until such time as the underpinning scientific research necessary for an accurate resource assessment is complete. Generally speaking, industry appears interested in the potential offered by SMS as they possess four major advantages over manganese nodules as a possible resource:

1. Average grades are some 1000 times more concentrated than those found in manganese nodules.
2. SMS tend to occur at approximately half the depth of manganese nodules (2500 m compared to 5000 m for nodules).

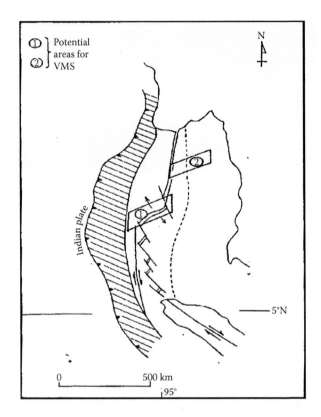

FIGURE 4.12

The potential areas for volcanogenic massive sulfide: (1) Unsedimented spreading axis, (2) sedimented spreading axis.

TABLE 4.5

Tectonic Setting with Example of SMS on the Modern Seafloor

1.	Sediment—Starved Mid-Oceanic Ridge?	
	Central Indian Ridge?	Southern Explorer Ridge
	Carlsberg Ridge	13°N and 21°N EPR
		Endeavor segment; N Juav deFuca Southern Juan de Fuca Ridge Galapagos Rift at 86°W, southern East Pacific Rise, TAG and Snakepit, MAR, Central Indian Ridge?
2.	Sedimented Mid-Ocean Ridge	Middle valley, Northern Juan de Fuca, Esianaba Trough, Southern Gorda Rift
3.	Seamounts	Axial seamount, Central Juan de Fuca East of 13°N EPR Green Seamount, West of 21°N EPR Tyrrnian Seamount, Mediterranean Sea
4.	Back Arc Oceanic Crust	
	Andaman Sea	Lau Basin Manus, Mariana Trough North Fiji Basin, Andaman Sea?
5.	Back Arc Continental Curst	Okinawa Trough North Aegnean Trough? Andaman Sea?
6.	Initial Rifting of Continental Margin	Guaymas Basins, Gulf of California, Atlantis II Deep, Red Sea, Western Woodlark Basin, Carlsberg Ridge?

3. Many SMS deposits are known to occur within EEZs whereas the major nodule deposits are all found in international waters, thus greatly simplifying rights of tenure for MPS in international law.

4. Processing of SMS ores is likely to be far less problematic than the equivalent for nodules.

Despite these apparent advantages over nodules, industry is extremely wary of committing hard cash to further investigations. This is in no small part due to the fact that hundreds of millions of dollars were spent on nodule research, which has still produced no commercial return.

A very strong consensus emerged from the survey, concerning the factors necessary to stimulate industry to take a more positive stance on SMS, such as applying for exploration leases and developing systems for India has national programs devoted to manganese nodule and SMS exploitation. Three major criteria need to be met before this is likely to occur; each of these will be discussed in more detail in later sections:

1. More data needs to be available concerning the distribution, grades, and general geology of SMS.

2. Metal markets must improve making extraction a more attractive commercial proposition.

3. A predictable and stable legal regime needs to be formalized, relating specifically to the extraction of these deposits.

In spite of the above constraints some companies from the United States, Europe, Japan, Canada, and others are taking a more positive interest in SMS.

4.7 The Scientific Significance of SMS Research

Economic benefit is by no means the only justification for studying sites of SMS formation. Active areas of deep-sea volcanic activity present a unique habitat, which has been colonized by a faunal community totally dependent on volcanic emissions for its existence. The deep sea was traditionally believed to be a desert merely supporting a sparse population of detrital feeders, which survived by scavenging the rain of organic material floating down from the euphotic above the water layers. This paradigm was overturned by the discovery on the Galapagos Ridge of a rich and diverse faunal community at a depth of 2600 m. Subsequent exploration has revealed many more such cases, each one centered around actively venting areas. The range of organisms new to science grows with each new find; giant tube worms up to 2 m long, blue octopi, giant mussels, and organisms so bizarre they warrant the creation of a new genus.

The basis of all this life are forms of bacteria, which are capable of utilizing the hydrogen sulfide, ubiquitous at emitting vents, to fix carbon in much the same way as green plants do on land. Strains of these bacteria have been found in a variety of vent organisms where they form a mutually beneficial symbiosis with their host, such as the giant tube worms, giant white clams, and a mussel. In each case, the actual mechanism of fixation varies as does the particular strain of bacteria, but essentially the host organism transfers hydrogen sulfide, oxygen, and carbon dioxide to the bacteria within its body, which then use the chemical energy to produce reduced carbon compounds, which are utilized as food by the host. This process is remarkable by the fact that hydrogen sulfide is extremely toxic, and each organism has also developed a means of protecting itself from its effects. These bizarre faunas in association with their "chemoautotrophic" bacteria provide the basis of a complex food chain, each member of which possesses mechanisms for protecting it from the potentially lethal sulfide-saturated vent environment.

The communities discovered around active vents are of immense interest to biologists and zoologists and they have also stimulated research into related areas. Similar communities of bacteria have now been found in such diverse environments as oil seeps, sewage outfall zones, marshes, and mangrove swamps. It is conceivable that such unique genetic material may prove to be of benefit to the biotechnologist. It has even been postulated that vent sites could be the original location for the evolution of life on the planet (Corliss et al., 1976).

Volcanic activity at spreading centers is more than a mere scientific curiosity. It has long been realized that this volcanism contributed large amounts of both heat and minerals into the ocean. The composition and temperature of the oceans, in turn, has a direct effect on the world's climate due to transfer of heat and gases, particularly carbon dioxide. Rona (1984) estimated that hydrothermal convection accounted for around 20% of the earth's total heat loss of $10.2 \times$ calories/s. Later, Edmond (1983) showed that the input of salts and metals into the oceans from ridge crest activity was equal in importance to fluvial sources, previously believed to be the only significant ones.

However, a recent discovery may indicate that these estimates of ridge crest contributions of heat and chemicals may be too low. In August 1986, a large hydrothermal plume was discovered over the Juan de Fuca Ridge off the Oregon coast. This "mega-plume," as it was dubbed, affected an area of seawater 19 km across and 0.7 km deep, the top of the plume coming to within 1300 m of the surface. Instruments towed through the plume indicated a 20-fold increase in some minerals and an average of 0.12°C rise in heat over ambient seawater. It was possible to show that the total heat released into this plume, 6.7×10^{16} J or 10 billion kilowatt hours of energy, was released in only a few days at most. This one mega-plume has produced the annual output of between 200 and 2000 high-temperature "black smokers" and so has radically altered current thinking about the likely contributions to the ocean from ridge crest activity. The cause of this phenomenon is believed to be a sudden release of hydrothermal fluids caused by a basaltic intrusion of approximately 0.01 km³. If this proves to be the case, then it is likely that hundreds to thousands of such plumes are be being generated each year along the global ridge system (Baker et al., 1987).

In addition to the outputs of heat and chemicals from the ridge the formation of SMS deposits appears to represent the primary ore-forming process on the planet. The initial concentrations of metals developed at spreading centers are carried along the "conveyor belt" of the seafloor and eventually will encounter a landmass. At this stage, they can be subjected to a variety of geological processes, each of which has the potential to give rise to further concentrations, which may become a commercial land-based ore body. Two examples will serve to illustrate of the possible chain of events, which in the geological past have led to the formation of thousands of ore-bodies now mined on land. The Troodos Complex in Cyprus has been mined since the Bronze Age for a variety of metals. Many of its ores are similar to those observed forming today at active spreading centers; some have even been found to contain what appear to be relicts of worm tubes commonly seen at sites such as the East Pacific Rise. The complex is, in fact, believed to be a section of oceanic crust, which millions of years ago was abducted onto an existing landmass by plate tectonic movement. The second example is copper porphyry deposits, currently the major source of the world's copper. These are found at convergent plate boundaries where oceanic crust (presumably containing relict SMS deposits) is thrust under a continental landmass causing it to melt. The resulting magmas make their way to the surface and solidify, and many have been found to contain localized concentrations of such metals as molybdenum or copper. It is likely that the original source of these concentrations was the relict SMS deposits contained in the oceanic crust. In both examples, it is possible that such land-based deposits may be subject to further geological processes, giving rise to second-ary deposits of ore grade material.

From the above discussion, it is evident that ancient SMS deposits have provided a cru-cial feedstock for many of the metalliferous deposits currently mined on land. As such, knowledge of their mode of formation and subsequent history can provide invaluable clues to the exploration geologist seeking land-based prospects anywhere in the world. Some studies have likewise shown a link between vent activity and the formation of other deep-sea deposits such as manganese nodules and co-rich crusts.

The scientific discoveries briefly outlined above coupled with the possibility of com-mercial gains from SMS deposits themselves have prompted many countries to launch scientific programs to investigate areas of undersea volcanism in more detail. The larger of these programs are reviewed in the following sections.

4.8 Future Trend

The seafloor massive sulfide industry is at a very early stage of mine development, but its activities are ahead of those of phosphorite, crusts, and perhaps nodules. There are only two serious commercial players in this field. Nautilus Minerals Inc. has extensive explo-ration leases in the Manus and Woodlark Basins territorial waters of eastern Papua New Guinea, in the EEZ of Solomon Islands and Tonga, and applications in New Zealand and Fiji. Neptune Minerals has exploration leases in the territorial waters of Papua New Guinea, and the EEZ of New Zealand, the Federated States of Micronesia and Vanuatu, and appli-cations in New Zealand, Japan, Commonwealth of Northern Mariana Islands, Palau, and Italy. Neptune has limited capital. Bluewater Metals, an unlisted private company based in Australia, has exploration licenses and applications in the southwest Pacific region to explore for seafloor massive sulfides. The partnership of Korea Ocean Research and Development Institute (KORDI) and Korea Institute of Geosciences and Mineral Resources (KIGAM), both government agencies, has exploration licenses from Tonga (Chosun Ilbo, April 3, 2008). Deep Ocean Resources Development Company (DORD), a Japanese private-government consortium under government control, has applied for concessions in the Japanese EEZ.

The key issue is the absence of legislation and regulation in many maritime states that would allow commercial exploration in their EEZ. Under UNCLOS, marine scientific researchers have the right, subject to the granting of clearance from the maritime state, to "explore," but commercial enterprises are unable to explore in many countries because the legislation and regulation to allow commercial exploration in the EEZ of these countries does not yet exist. This is not a level playing field for exploration and can create conflict if the maritime state believes that marine scientific researchers are fronts for companies. Without a wider range of tenements, it will be more difficult for industry to advance. Exploring several tenements in different geological provinces increases the chances of success, especially given the offshore attribute of aggregate deposits by moving the floating infrastructure. Exploration is the only way to gain knowledge. Exploration does not imply mining will be allowed; it is simply needed to gather information that may have a wide variety of applications.

Another issue is the development of appropriate ISA regulations for exploiting "polymetallic sulfides" in international waters. Many Pacific island nations will likely adopt these regulations or parts thereof, and are waiting to see them. Thus the impact of the proposed ISA regulations for "polymetallic sulfides" will be far greater than for waters beyond national jurisdictions. Consequently, the ISA must have input from stakeholders, especially from commercial enterprises such as mining companies, marine contractors, and environmental contractors who would likely carry out the work of preparing an environmental impact assessment. Earlier ISA draft regulations for "polymetallic sulfides" appeared to be largely based on inputs from scientific stakeholders who have a poor understanding of business fiscal reality. The ISA has now solicited input from industry and has incorporated a number of its suggestions.

References

Anon, 1981. Andaman Sea—Gulf of Thailand (Transect IT). Studies in East Asian Tectonics and Resources (SEATAEATAEATAR), CCOP, Bangkok: p. 37–51.

Baker, E. T., Massoth, G. J., and Feely, R. A., 1987. Cataclysmic hydrothermal venting on the Juan de Fuca ridge. *Nature*, 329: 149–151.

Banerjee, P. K., Saha, B. K., Majumdar, S., Rakshit, S., Deb Roy, D. K., Sinha, J. K., and Bhattacharya, D., 1992. Nature and origin of anomalous base metal enrichments in the recent sediments around the Narcondam Island in the Andaman Sea. In: *Recent Geoscientific Studies in the Bay of Bengal and the Andaman Sea*, Geological Survey of India, Special Publication, Kolkata, No. 29, pp. 111–116.

Banerjee, R. and Ray, D., 2003. Metallogenesis along the Indian Ocean ridge system. *Curr. Sci.*, 85: 321–327.

Bignell, R. D., 1978. Genesis of red sea metalliferous sediments. *Mar. Min.*, 212–220.

Biswas, S., Majumdar, R. K., and Das Gupta, A., 1992. Distribution of stress axes orientation in the Andaman-Nicobar Islands region: A possible stress model and its significance for extensional tectonics of the Andaman Sea. *Phys. Earth Planet In.*, 70: 57–63.

Bostrom, K., and Peterson, M. N. A., 1966. Precipitates from hydrothermal exhalations on the East Pacific rise. *Econ. Geol.*, 61: 1258–1265.

Cann, J. R., Winter, C. K., and Pritchard R. G., 1977. A hydrothermal deposit from the floor of the Gulf of Aden. *Mineral. Mag.*, 41: 193–199.

Corliss, J. B., Dymond, J., Lyle, M., Doerge, T., Crane K., Lonsdale, P., Von Herzen, R. P., and Williams, D., 1976. Sediments mound ridges of hydrothermal origin along the Galapagos rift [abs]: *EOS, Am. Geophys. Union Trans.*, 57: 935.

Curray, J. R., 1991. Possible green schist metamorphism at the base of 22 km sedimentary section, Bay of Bengal. *Geol. South Pacific Geol. Notes*, 19: 1097–1100.

Curray, J. R., Moore, D. G., Lawver, L. A., Emmel, F. J., Raitt, R. W., Henry, M., and Kieckhefer, R., 1979. Tectonics of the Andaman Sea and Burma. *Am. Assoc. Petrol. Geologists: Memoirs*, 29: 189–198.

Curray, J. R., Moore, D. G., Lawver, L. A., Emmel, F. J., Raitt, R. W., Henry, M., and Kieckhefer, R., 1982. Structure, tectonics and geological history of the North-eastern Indian Ocean. In: A. E. M. Nairn and F. G. Stehli (eds.). *The Ocean Basin and Margins: The Indian Ocean*. Plenum Press, New York, Vol. 6, pp. 399–450.

Degens, E. T. and Ross, D. A. (eds.). 1969. *Hot Brines and Recent Heavy Metal Deposits in the Red Sea*. Springer-Verlag, New York, p. 600.

Degens, E. T. and Ross, A. R., 1970. The red sea hot brines. *Science* 222(1970): 32–42.

Dutta, S., 1991. Barren Island volcano spits fire. *Geological Survey of India*, Marine Wing Newsletter 7(2): 3–5.

Edmond, J. M., 1983. Chemistry of the 350°C hot springs on the crest of the EPR at 21°N. *J. Geochem. Explor.*, 19: 491–492.

Eguchi, T., Uyeda, S., and Maki, I., 1979. Seismotectonics and tectonics history of the Andaman sea. *Technophys*, 57: 35–51.

Fujimota, H., Cannat, M., Funioka, K., Gamo, T., German, C. R., Mevel, C., Munch, U., Ohta, S., Oyaizu, M., Parson, L., Searle, R., Sohrin, Y., and Tama-ashi, T., 1999. First submersible investigations of mid-ocean ridges in the Indian Ocean. *InterRidge News* 8(a): 22–24.

Fouquet, Y., Stackelberg, U. von, Charlou, J. L., Donwal, J. P., Foucher, J. P., Erzinger, J., Herzig, P., Muhe, R., Weidicke, M., Soakai, S., and Whitechurch H., 1991. Hydrothermal activity in the lau back-arc basin: Sulphides and water chemistry. *Geology*, 19: 303–306.

Gamo, T., Chiba, H., Yamanaka, T., Okudaira, T., Hashimoto, J., Tsuchida, S., Ishibashi, J.-I., Kataoka, S., Tsunogai, U., Okamura, K., Sano, Y., and Shinjo, R., 2001. Chemical characteristics of newly discovered black smoker fluids and associated hydrothermal plumes at the Rodriguetz Triple Junction Central Indian Ridge. *Earth Planet. Sci. Lett.* 193: 371–379.

German, C. R., 2003. Hydrothermal activity on the eastern southwest Indian Ridge (50–70°E): Evidence from core-top geochemistry, 1887 and 1998. *Geochem. Geophys. Geosyst.*, 4: 9103. http://dx.doi.org/10.1029/2003GC000522.

Guennoc, P. and Thisse, Y., 1982. Genesis of the opening of the red sea and the mineralizations of the axial rifts: Bureau de recherché geologiques et minieres [Paris] documents, 51, 144.

Halbach, P., Blum, N., Munch, K., Pluger, W., Garbe-Schonberg, D., and Zimmer, M., 1998. Formation and decay of a modern marine sulfide deposit in the Indian Ocean. *Mineralium Deposita*, 33: 302–309.

Halbach, P., Blum, N., Munch, U., Pluger, W., and Kunh, T., 1996. The Sonne sulfide field is not alone in the Indian Ocean. *BRIDGE Newslett.*, 10: 51–54.

Halbach, P., Blum, N., and Pluger, W., 1994. The Sonne Field: The First Massive Sulphides from the Indian Ocean floor, Abstract, 25th Underwater Mining Institute Conference, Montery Bay (California).

Halbach, P., Nakamura K., Wahsner, M., Lange, J., Sakai, H., Kaselitz, L., Hansen, R. D., Tamano, M., Post, J., Seifert, R., Michaelis, W., Teichmann, F., Kinoshita, M., Marten, A., Ishibashi, J., Czerwenski, S., and Blum, N., 1989. Probable modern analogue of kuroko-type massive sulphide deposits in the Okinawa Trough Back-arc Basin. *Nature*, 338: 496–499.

Halder, D., Laskar, T., Bandopadhyay, P. C., Sarkar, N. K., and Biswas, J. K., 1992. Volcanic eruption of the Barren Island Volcano, Andaman Sea. *J. Geol. Soc. India*, 30: 411–419.

Haymon, R. and Kastner, M., 1981. Host springs deposits on the East Pacific Rise at 21 deg N. Preliminary description of mineralogy and genesis. Earth Planet. Sci. Letts., 53: 363–381.

Herzig, P. M. and Plüger, W. L., 1988. Exploration for hydrothermal activity near the Rodriguez Triple Junction, Indian Ocean. *Can. Mineral*. 26: 721–736.

Jian, Z., Jian, L., ShiQuin, G., and YoungShun, C., 2008. Hydrothermal plume anomalies along the Central Indian Ridge. *Chin. Sci. Bull.* 53: 2527–2535.

Kumagai, H., Nakamura, K., Toki, T., Morishita, T., Okino, K., Ishibashi, J.-I, Tsunogai, U. et al. 2008. Geological background of the Kairei and Edmond hydrothermal fields along the Central Indian Ridge: implications of their vent fluids' distinct chemistry. *Geofluids*, 8: 239–251.

Mukhopadhyay, M., 1984. Seismotectonics of subduction and backarc rifting under the Andaman Sea. *Tectonophysics*, 108: 229–239.

Mukhopadhyay, M., 1988. Gravity anomalies and the deep structure of the Andaman Arc. *Mar. Geophys. Res.*, 9: 197–210.

Munch, U., Halbach, P., and Fujimoto, H., 2000. Sea-floor hydrothermal mineralization from the Mt. Jourdanne, Southwest Indian Ridge. *JAMSTEC J. Deep Sea Res.* 16: 125–132.

Pandey, J., Agarwal, R. P., Dave, A., Maithani, A., Trivedi, K. B., Srivastava, A. K. and Sugh, D. N., 1992. *Geol. Andaman Bull, O.N.G.C.*, 29: 19–103.

Pluger, W. L., Herzig, P. M., Becker, K. P., Deismann, G., Schops, D., Langer, J., Jensich, A., Ladage S., Richnow, H. H., Schulzee, T., and Michaelis, W., 1990. Discovery of hydrothermal field at the Central Indian ridge. *Mar. Min.*, 9: 73–86.

Rodolfo, K. S., 1969. Bathymetry and marine geology of the Andaman Basin: Tectonic implications for Southeast Asia. *Geol. Soc. America Bull.*, 80: 1203–1230.

Rona, P. A., 1984. Hydrothermal mineralization at seafloor spreading centers. *Earth-Sci. Rev.*, 20: 1–104.

Rona, P. A., 1988. Hydrothermal mineralization at oceanic ridges. *Can. Min.*, 26: 431–465.

Roonwal, G. S., 1986. *The Indian Ocean: Exploitable Mineral and Petroleum Resources.* Springer-Verlag, Heidelberg, Vol. XVI, p. 198.

Roonwal, G. S., 1997. Marine mineral potential in India's exclusive economic zone: Some issues before exploitation. *Mar. Georesourc. Geotechnol.*, 15: 21–32.

Roonwal, G. S., 2005. Mineral endowment of the Indian Ocean. In: Harsh K. Gupta (ed.). *Oceanology.* Universities Press, Hyderabad, pp. 75–91.

Roonwal, G. S., 2009. Sea bed Sulphides, Gas Hydrates, and Hydrocarbon Resources in the Andaman Sea: North East Indian Ocean, *Proceedings of the 8th ISOPE Ocean Mining Symposium*, pp. 96–101.

Roonwal, G. S. and Glasby, G. P., 1996. Co-rich manganese crusts and hydrothermal mineralisation the Indian Exclusive Economic Zone. In: S. Z. Qasim and G. S. Roonwal (ed.). *India's Exclusive Economic Zone.* Omega Scientific Publishers, New Delhi, pp. 58–64, 238 pp.

Roonwal, G. S., Glasby, G. P., and Chugh, R., 1997a. Mineralogy and geochemistry of the surface sediments from the Bengal Fan, Indian Ocean. *J. Asian Earth Sci.*, 15: 33–41.

Roonwal, G. S., Glasby, G. P., and Roelandts, I., 1997b. Mineralogy and geochemistry of surface sediments from the Andaman Sea and Upper Nicobar Fan, Northeast Indian Ocean. *J. Indian Assoc. Sedimentol.*, 16(1): 89–101.

Roonwal, G. S. and Mitra A., 1988. Hydrothermal sulphide mineralization on the sea floor spreading centers: A review. *Indian J. Mar. Sci.*, 17: 249–257.

Rozanova, T. V. and Baturin, G. H., 1971. Hydrothermal ore shows on the floor of the Indian Ocean. *Oceanology*, 11: 874–879.

Wang, T. T., Chen, Y. S. J., and Tao, C. H., 2011. Revisit the K-segment of the southeast Indian ridge for new evidence of hydrothermal plumes. *Chin. Sci. Bull.*, 56: 3605–3609. http://dx.doi.org/10.1007/s11434-011-4723-5.

5

Shallow Oceanic and Coastal Mineral Deposits

5.1 General

The coastal zone, as a resource in its own right, is one of our most precious geographies because much of the world population lives within the coastal zone. These zones are engaged in a range of activities such as food production (agriculture), commercial cultivation, business, and factories. All this human activity places a big stress on the fragile coastal zone, which responds adversely to change. For example, the development of a harbor or dredging of a channel in one area may induce erosion in another. Building a dam can affect the input of sediments or nutrients, leading to drastic change to both living and nonliving resources of the near-shore zone. Natural phenomena such as an earthquake or flood can result in major changes to the shoreline; some of the changes can be global in its reach such as the rise of sea level as a result of global warming, which will have a major effect on coastal zone including erosion, salinity increase, and the destruction of coastal aquifer system.

5.2 Phosphorite

Sedimentary phosphate deposits are an example of shallow marine mineral deposits that form part in response to physical, chemical, and biochemical changes within the water column and at the sediment–water interface. However, they are not solely shallow marine. They have been known since the HMS *Challenger* expedition to occur as crust and nodules on plateau and seamounts.

Sedimentary phosphate deposits are the major source of phosphate for phosphatic fertilizers. Without phosphatic fertilizers, it would be impossible to sustain world food production to meet the requirements of rising population. For this reason, phosphate is a major world commodity with a total world production of about 210 million tonnes of phosphate rock with the value of several billion dollars. Though most of this production happens in North America, North Africa, and the CIS, India produces 14 million tonnes of rock phosphate from mines in Udaipur, Jhabhua, and Mussoorie. Looking at the world scene, there is no shortage of phosphate resources, but rather uneven distribution of those resources gives rise to abundant and deficient regions. India's production of rock phosphate is not sufficient to meet the domestic demand. Therefore, the country imports phosphate from Morocco and other West Asian countries. Marine phosphate will play an important role to supplement our demand.

There are various types of phosphorite, but by far the important type is composed of well-rounded size or pellets. Phosphorite is considered to form under relatively shallow marine conditions in areas of high organic productivity as the organic remains accumulate; they produce organic-rich sediment but not necessarily on phosphorite. Formation of phosphorite requires processes such as removal of organic matter, digenetic phosphatization, mechanical reworking, and so on. Surely the actual processes involved are by far complex than this. Offshore phosphatic occurrences have been known in the mud banks of Malabar Coast. Surveys by the Geological Survey of India, as well as by the National Institute of Oceanography, have identified phosphatic sediments off Saurashtra coast in the west and Coromandel coast on the east.

Phosphorite, a fertilizer mineral, is primarily a calcium phosphate. Therefore, as compared to other sedimentary rocks, it is distinguished by high phosphorite (P_2O_5) content. A phosphorite rock or ore can have a variety of mineral apatite, which may vary between 5% and as high as even 40% (Baturin, 1982; Kudrass, 2000).

Actually, phosphorite deposits are a part of a biogeochemical cycle. The source of element is the erosion by terrestrial rocks, and on erosion, it gets transported through rivers to the sea. It is then a procedure of uptake of phosphorus by plankton. It gets accumulated in the shallow marine water, resulting due to sinking and dissolution of planktons. It is then pushed to surface water due the process of upwelling seen clearly in the Indian Ocean, along the west coast of Indian peninsula along the horn of Africa (Socotra island) in the South African coast.

In the marine environment, phosphorite may occur in the following settings (Hein, 2008):

1. On the continental shelves and ships, commonly seen along west coast of Indian peninsula, due to effect of upwelling and trade winds.
2. On the submarine plateau, ridge and banks as seen as seen in Chatham Rise off New Zealand.
3. As island atolls, generally within atoll lagoons.
4. Possibly on midoceanic plate seamounts, resulting from replacement of carbonates by fluorapatite.

5.3 Construction Material

Sand and gravel are extracted from many offshore areas, particularly adjacent to highly populated areas of western Europe, east coast of the United States, and Asia. The sand and gravel was for the most part deposited during a previous low sea-level strand than subsequently inundated by the postglacial sea-level rise. Calcium carbonate (calcareous) sand is an important mineral resource and can be mined from several coastal areas where there are accumulations of shells, calcareous sands, or even dead coastal reefs. India has yet to begin such mining operation on a commercial basis.

Lime mud deposits have been delineated over a length of 80 km in the water depth of 102–200 m off Godavari delta sediments. The average thickness varies between 1 and 2 m, and it is overlain by a layer of green clay, which suggests hiatus in the sedimentation. This lime is a creamy white, and Ca content varies between 46% and 51% (Rao and Reddy, 1993; Rao et al., 1992). Lime mud has also been located off the Gujarat coast in a water depth of

180–200 m and occupies an area of 3200 km^2. At the shelf-edge width of 40 m, $CaCO_3$ content is 92%–94% but goes down to 72%–75% in the upper and lower slope area (Vaz et al., 1993).

Calcium mud occurs in Laccadive Sea. These are primarily coral atolls, submerged reefs, and banks and were surveyed by the Geological Survey of India (Siddique and Mallik, 1973). These are good quality with a CaO content of 50%. Because they are low in silica content, the economic significance increases and an estimate of 1 million tonnes in a water depth of 280 m is expected.

5.4 Heavy Mineral Placers

Beach heavy minerals or placers are formed by mechanical erosion of terrestrial rocks, transported, sorted, and accumulated by flowing water. They result in placer deposits in beach and offshore sediments of continental margins because of their higher density (>3.2 g cm^{-3}) relative to the bulk of detrital minerals consisting mostly of quartz and feldspar (2.5–2.7 g cm^{-3}). Due to their hardness, they are resistant to weathering and mechanical action during transport determines the distance it can be transported from its source without material change of state (Kudrass, 2000). The median distance of transport from a bedrock source to an offshore placer deposit is 8 km (Emery and Noakes, 1968). As noted, an outstanding feature of the distribution of placer deposits is the multitude of coastal sites known and the few of these sites of past or present mining (Table 5.1).

Three generic types of placer deposits are recognized (Kudrass, 2000; Rona, 2008): There are (i) disseminated beach placers generally comprising light heavy minerals (density <6 g cm^{-3}; e.g., rutile for Ti; ilmenite for Ti; magnetite for Fe and REE; monazite for Ce, La, Y, and Th; zircon for Zr; sillimanite and garnet), which are concentrated by waves and long-shore currents; (ii) drowned fluviatile placers comprising coarse sand and gravel overlying the bottom of river channels containing heavy metals (e.g., cassiterite for Sn, gold); and (iii) eluvial deposits containing heavy metals. Placer deposits may lie above, at, and below present sea level related to the history of regional and eustatic sea-level change. In the geologic record, fluviatile placers are the presently most important from an economic point of view. Gold, tin, and diamond placer deposits are important placer deposits.

Heavy minerals occur at several locations in the Indian Ocean (Figure 5.1). Marine placers form in response to present-day conditions or are residual deposits left from a previous energy regime. Mineral sands containing economically significant minerals such as rutile, ilmenite, zircon, and monazite occur in the present-day beaches in various parts of the world such as Australia, southeastern United States, West Africa, and peninsular India. However, mineral sands also occur in old beach systems generally landward but occasionally seaward of the present-day shoreline. Therefore, obviously knowledge of relative sea-level changes should assist the exploration of heavy mineral sands. Many placer deposits such as the offshore tin deposits of Southeast Asia are found in old river channels that have been inundated by the postglacial rise in sea level. Diamonds have been mined from the beaches and immediately offshore dredging operations in Canada, Alaska, West Africa, and the Philippines. In India, the placer deposits are mined on the southwest coast of Kerala and in the east in Odisha. These deposits are famous for ilmenite, rutile, zircon, and monazite. On the east coast of India, similar types of deposits occur on the beaches of Tamil Nadu, Andhra Pradesh, and Odisha. Some details of these follow.

TABLE 5.1

Distribution of Placers and Exploration Status along Indian Coasts

		Placers Available	Status of Exploration
West Coast	Gujarat	Mainly ilmenite and magnetite	Onshore and offshore occurrences reported. Still exploration needed.
	Maharashtra	Ilmenite and magnetite mainly	Onshore and offshore exploration completed. Appreciable reserves of ilmenite proved.
	Goa	Ilmenite, magnetite, and zircon	Onshore surficial occurrences reported. Systematic exploration is lacking for offshore. Onshore sampling completed. Exploration being conducted.
	Karnataka	Ilmenite, magnetite, garnet, etc.	Onshore surficial occurrences reported, exploration continued.
	Kerala	Ilmenite, magnetite, rutile, zircon, monazite, sillimanite	Detailed on shore, offshore exploration completed. Beach placers being exploited at Chavara.
	Tamil Nadu (West Coast)	Ilmenite, magnetite, zircon, rutile, sillimanite	Onshore and offshore exploration completed. Placers being exploited at Manavalakurichi
East Coast	Tamil Nadu (East)	Ilmenite, magnetite, zircon, garnet, rutile	Onshore exploration completed. Some deposits of garnet and ilmenite being mined. Work needed in offshore.
	Andhra Pradesh	Ilmenite, magnetite, etc.	Onshore and offshore exploration completed
	Odisha	Ilmenite, monazite, zircon, rutile, etc.	Onshore and offshore exploration completed. Exploitation at Chhatrapur, Gopalpur being continued.
	West Bengal	No systematic data	Exploration needed.

Source: After Mallik, T. K., 2007. *Indian Geol. Congr. Bull.*, I: 82–87.

5.4.1 Distribution of Heavy Mineral Sand along the Indian Coast

Good deposits of heavy mineral beach sand occur both on the east and west coast of the country (Figure 5.1 and seen from Table 5.3) (Table 5.2) (Atomic Minerals Directorate, 2001).

Review on beach placers in India includes Rajamanickam (1996) on beach and placer deposits of Neendakara–Kayamkulam in Kollam district, Kerala, and Manavalakurichi in Kanyakumari district, Tamilnadu, and Chhatrapur, Odisha; Mahadevan and Sriramdas (1948) on black sand concentration along the Visakhapatnam–Bhimunipatnam coast, Andhra Pradesh; Rao et al. (1992) on southern parts of Andhra coast between Ramayapatnam (Nellore) and Vashista Godavari (Kakinada); Rao et al. (1992) on the heavy mineral concentration and reserve of deposits between Lawson's Bay and Bhimunipatnam–Konada (north of Visakhapatnam) in Andhra Pradesh; and Ali et al. (2001, 1998) on the placer deposits of Ratnagiri district, Maharashtra. Exploration carried out by AMD in the past five decades has resulted in identifying heavy mineral placer deposits along the coastal stretches of Maharashtra, Kerala, Tamil Nadu (including inland placers), Andhra Pradesh, and Odisha (Figure 5.1).

The economic interest in titanium concern the oxides of titanium. The silicate mineral of titanium like sphene (CaO · TiSiO$_4$), regardless of its titanium contents (about 41%), cannot be used as an ore of titanium. Technology is yet to be developed even for the oxide mineral

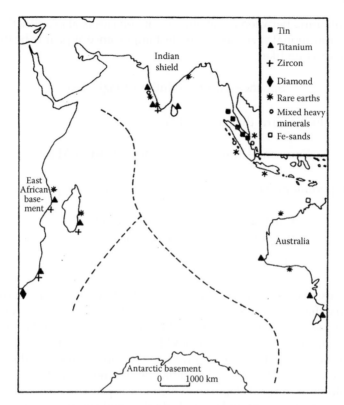

FIGURE 5.1
Major distribution of heavy minerals in the Indian Ocean.

(perovskite, $CaO \cdot TiO_2$) that contains as high as 59% TiO_2. Due to this, the only minerals that have economic significance are ilmenite ($FeO \cdot TiO_2$), rutile (TiO_2), and anatase (TiO_2). During the alteration process, ilmenite changes to leucoxene during which the TiO_2 content increases to as high as 70% with concomitant removal of FeO.

Titanium is usually immobile during metamorphic processes with its content in the rocks generally remaining more or less constant. However, quantity and separation of titanium in oxide and silicate phases vary greatly among metamorphic facies. To cite an example, 1% TiO_2 may be present as 2.5% sphene at low-grade metamorphism or as 2.0% ilmenite or 1% rutile at higher grades of metamorphism.

TABLE 5.2

Total Indian Reserves of Beach Sand Placer Resources

Mineral	Resource in Million Tonnes
Ilmenite	348
Rutile	18
Zircon	21
Sillimanite	130
Garnet	107
Monazite	8

Titanium partitioning between oxide and silicate phases is also a function of rock composition. The compositional variables of greatest importance appear to be the Al/Ca value, which governs a reaction like

$$Sphene + Ai_2O_3 \rightarrow Rutile + plagioclase$$

whereas Fe/Mg value governs the reaction like

$$Pyroxene + Rutile \rightarrow Limonite + Quartz + MgO$$

Some other important possible reactions during metamorphic processes are

$$Ilmenite + O_2 \rightarrow Magnetite + Rutile$$
$$Ilmenite + S_2 \rightarrow Pyrite + Rutile$$
$$Sphene + CO_2 \rightarrow Rutile + Calcite + Quartz$$

India's resources base is sufficiently large for it to become a more significant force in titanium industry than it is at present. The Australian TZ Mineral International Property Limited of Perth gave an excellent profile of Indian mineral sand industry including historical background, current development, and likely future trend.

In general, rocks of low-grade metamorphism contain titanium in the form of sphene and also to some extent in biotite and hornblende. We notice that two suites of igneous rocks contain titanium mineral deposits, which are at present considered valuable: first, anorthosite–ferrodiorite massifs and second, alkaline suite of rocks. In the anorthosite-ferrodiorites, it is the ferrodiorite that are enriched in TiO_2 and contain titanium oxide minerals in abundance. The bodies of ferrodiorite to ferrogabbro significantly contain more TiO_2 and P_2O_5 content than some other rocks such as diorites and gabbros. Their total iron oxide content relative to MgO is also unusually high.

These rocks occur as dykes intruding the Precambrian older anorthosites (1700–1900 Ma) or other country rocks sequence. They contain ilmenite, magnetite, hematite, spinel, and minor rutile. The presence or absence of ilmenite as individual single crystals relatively free of intergrowth decides the economic value of a deposit. In general, TiO_2 present in magnetite is valueless. About 40% of titanium mineral resources is being mined from hard rocks of igneous origin and the rest 60% comes from placer sands. Some good examples of titanium mineral deposits from igneous rocks are Allard lake (Quebec, Canada), Tallnes, and Sorgangan deposits of southern coast of Norway and Roseland district (Virginia), San Gabriet range (California), Laramic range (Wyoming), and Duluth complex (Minnesota) (all from the United States). The Roseland district (Virginia) stands as a good example for both contact metasomatic rutile and magmatic ilmenite deposit.

The other important source for titanium mineral deposits is within alkaline suite of rocks. Zircon ($ZrSiO_4$) is the chief source of zirconium. Because of higher charge, that is, 4, and ionic radius of 0.59 Å, Zr is mostly confined to the granites and pegmatite in the form of the mineral, zircon. Zircon is highly resistant to physical and chemical weathering and, as a result, zirconium, in the form of mineral zircon, gets concentrated in placer sands. World zirconium resources come from zircon placer sands and nonzircon deposit has so far been mined from other sources.

Baddeleyite (ZrO_2) occurs as a gemstone in pegmatite and also in some placer deposits in Sri Lanka and Brazil.

The rare earth thorium is the most important and the common mineral consisting of both rare earths and thorium with minor amounts of uranium is monazite. In igneous rocks, monazite is characteristically present in highly differentiated suites like granites and pegmatite. The thorium content in monazite ranges from negligible amount to 32% ThO_2 (in cheralite) and that of rare earths is 55%. With higher grade of metamorphism, the Th content of monazite increases. High-grade (granulite facies) metapelitic rocks contain ore monazite than the low-grade green-schist facies.

Sillimanite (Al_2SiO_5) and kyanite (Al_2SiO_5) are characteristic minerals of metamorphism, mainly in politic rocks. Sillimanite is formed in high-grade metamorphic rocks of granulite facies and is in close association with garnet (almandine and spessartine variety with minor proportion of pyrope). Kyanite is characteristic of medium-grade metamorphism with intermediate temperature and high pressure.

Garnet is a common placer mineral both in riverine and beach environments. It occurs in medium- to high-grade metamorphic rocks (almandine in upper amphibolites and granulite facies and pyrope in eclogite facies), calc-silicate rocks (scarns) in the form of grossularite and spessartine-almandine variety in pegmatites. Pyrope variety of garnet may also occur in some ultramafic rocks. Para-gneisses like khondalite (garnetiferous sillimanite gneiss) and a few other rocks like leptynite are characteristically garnetiferous.

5.4.1.1 West Coast of India

The west coast of India has one of the rich placer deposits of ilmenite, rutile, and monazite in the Malabar Coast of Kerala and Ratnagiri–Konkan coast of Maharashtra where magnetite–ilmenite placers occur in economic concentration.

Somewhat detailed survey and sampling has been conducted in region near Kollam district (earlier Quilon) where in Chavara and Varkala sections rich association of ilmenite, sillimanite, zircon, rutile, and monazite, leucoxene, garnet, kyanite, and ferromagnesian minerals occur. Of them, the working unit at Chavara by the Indian Rare Earths Ltd. and by Kerala State Minerals Corporation are profit-making mining units. Bulk of the produce is sold to Travancore Paint, the Atomic Energy Commission, and spare material is exported (Rao, 1988).

Beneficiation study by the Indian Rare Earth Ltd., Chavara, regularly monitors the quality of placers mined. On average, the Chavara samples have an assemblage of 11% ilmenite, 3% rutile, 3% zircon, 0.2% monazite, and 2.4% sillimanite. Vibrocover sampling in Varkala area have shown similar results. The Konkan–Ratnagiri ilmenite magnetite deposits have a good tenor and could be exploited in the course of time.

5.4.1.2 East Coast of India

Already surveyed and known areas are in Tamil Nadu, Andhra Pradesh, and Odisha coast. Of them, the Odisha coast has good deposit in Chhatrapur, Gopalpur, where the Indian Rare Earths Ltd. has a commercial plant (Figure 5.1). Heavy mineral association is similar to those of west coast (near Chavara), but the grade is low. However, the deposits occur on larger tracts than in the Quilon coast in Kerala state. In Andhra Pradesh, new deposits have been located northward between Sonapur and Mahanadi estuary in Odisha. They appear as extension of Gopalpur deposits and have its origin to the Eastern Ghat khondalite rocks. Their bulk composition is given in Table 5.3.

India's resources base in heavy mineral sand (Figure 5.1) is sufficiently large for it to become a far more significant force in rare earths and TiO_2 industry than it is at present; new development and open for policy may change this situation.

TABLE 5.3

Total Heavy Mineral Wise Placer Concentration as
wt.% in Chhatrapur (Gopalpur), Odisha

Minerals	wt.% of Heavy Mineral
HM concentrate—bulk	9.3
Ilmenite	4.2
Sillimanite	1.6
Garnet	2.8
Monazite + zircon + rutile	0.41
Others	0.30

A review on the mineral sands of India can be obtained through TZM1 (2000), India's Atomic Minerals Directorate (2001) special issue on beach and inland heavy mineral sand, in addition to vast information in the various reports of the Geological Survey of India.

The world's first heavy mineral sand production plant was established in India 1911 at Manavalakurichi in Tamilnadu on the southern tip of India, based on large coastal deposits of monazite discovered two years earlier (1909). The first ilmenite shipment was made in 1922, and in the following decade, the output grew rapidly with the development of Chavara deposits near Kollam (Quilon) in the adjacent southern state of Kerala. According to TZM1 (2000), India was the world's largest ilmenite exporter during the 1930s and 1940s operating four plants and exporting 300,000 t (in 1940). The industry subsequently went into decline and by 1965, three of the four plants had closed. Exports were affected by two factors: the unsustainability of the Kollam ilmenite for pigment production using the sulfate route and the passing of content of ilmenite export to Atomic Energy Commission (AEC) of India, which was established in 1947 at the time of independence. Under the atomic mineral legislation enacted at that time, the AEC took control of all "atomic minerals." These depend as those containing in excess of 0.25% monazite, and thus private sector investment in mineral sands was, in effect, stifled.

The current mineral sand industry is dominated by Indian Rare Earth (IRE), which accounts for 95% of the output. IRE was established in 1965 as a Central government undertaking and acquired the original operation at Chavara and subsequently the operation at Manavalakurichi. Indian titanium industry is relatively small on a global scale, but highly diverse. There are four mining and processing sites in Tamil Nadu and the northeastern state of Odisha. Collectively, they have an ilmenite production capacity of some 400,000 t/y. During the year 2010, the output was around 400,000 t, with IRE's Odisha Sands Complex (OSCOM), operating in Odisha, contributes 185,000 t (Figures 5.2 and 5.3). India also possesses four synthetic rutile plants with a combined capacity of some 90,000 t. The production year 2016 was about 50,000 t/y; in addition, there are four TiO$_2$ pigment plants, including one in Kolkata (Calcutta) in West Bengal. They have a combined production of about 45,000 t.

Domestic consumption of ilmenite since 1999 has been lower as a result of lower offtake (90,000 t) by the pigment industry. Demand for ilmenite for synthetic rutile showed little change at 42,000 t. TZM1 estimates total exports of ilmenite is a little over 150,000 t, shipped mainly to Europe, with smaller volumes detained for the United States and South Korea. India has yet to capture market in Japan, which at present imports REEs from China. Of Indian total synthetic rutile output of 50,000 t, only 15,000 t were exported compared with 30,000 t in the previous years. Japan appears to be the only destination. Domestic usage of synthetic rutile is estimated at 20,000 t.

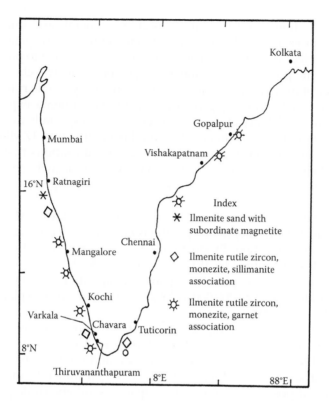

FIGURE 5.2
Beach heavy mineral sand deposits in India.

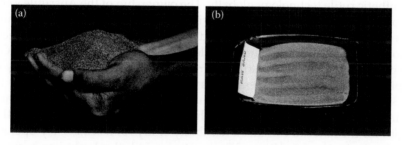

FIGURE 5.3
(a) Rutile rich heavy mineral sands on Odisha coast and (b) sillimanite rich heavy mineral sands on Odisha coast.

India's ilmenite resources base, as identified by the Atomic Energy Department, is reportedly as high as 200 mt, and the ilmenite quality varies between 50% and 60% TiO_2.

5.5 Economic Consideration and Application of Rare Earths in High Technology

In October 1998, a change of government policy paved the way for the select entry of Indian and foreign private investors. TZMI reports that participation of wholly owned

Indian companies is now permitted in all facets of mineral-sand mining, separation, and value-adding projects and that upto 70% foreign equity participation is possible for value-adding and integrated projects. The policy appears designed to preclude the export of ilmenite unless the project has a value-adding component. The change in policy has spurred a number of initiatives, with a principal focus on synthetic rutile. TZMI further points out that this is not surprising given the constraint on ilmenite exports and the high costs and uncertain availability of electric power in India, especially during the summer months, which makes the production of titanium slag unattractive. TZMI provides details about the number of new developments in the offing but cautious that they should be viewed at best as being under active investigation. This is because the combination of raising adequate finance and securing the necessary government approval has delayed their confirmation.

In Kerala, the government owns Kerala Mineral and Metals Ltd. (KMML), and the company already conducts fully integrated operations ranging from mining to synthetic rutile output and pigment production. Production is based on reserves of 60% TiO_2, and the same deposit at Chavara, mined by IRE, is exported. KMML's current ilmenite production is 200,000 t/y, but there are plans to expand the capacity to 330,000 t/y, and the raise synthetic rutile production from 200,000 to 230,000 t/y. There are also plans to lift pigment output from 22,000 to 60,000 t/y.

In Odisha, there is an interesting development involving Austpac Resources NL of Australia and IRE. The Australian rutile joint venture was formed in August 1999 to develop Auspac's ERMS synthetic rutile technology, and the first objective is to build a 10,000 t/y demonstration plant adjacent to IRE's OSCOM mineral sands operation near Chhatrapur. OSCOM possesses a 100,000 t/y capacity synthetic rutile plant, but it is not currently in production. In 2006, Ticor Ltd. announced its decision to participate in the Ausrutile joint venture. Ticor Ltd. is already in a joint venture with Austpac to develop and take the synthetic rutile technology globally. The ownership of the Ausrutile joint venture will now comprise Austpac and Ticor each with 37%, and IRE holding the remaining 20%.

Elsewhere in Odisha, a privately owned Indian company, Saraf Agencies, is also reported to be considerably a synthetic rutile plant based close to IRE's OSCOM operations. The plant, as envisaged, would be a first in India. A 100,000 t/y capacity operation is contemplated using OSCOM ilmenite as a feedstock, with the grade as low 50% TiO_2. However, Saraf has also applied to the Odisha government for rights to mine an ilmenite deposit lying beyond IRE's lease area and has signed an agreement with MD Mineral Technologies for the latter to conduct a feasibility study for a 500,000 t/y operations. MD Mineral Technologies is a subsidiary of Evans-Deakin Ltd., the Australian heavy engineering group and is a leading designer and supplier of processing equipment for the mineral sand industry.

Asia has one of the largest reserves of rare earth ores. It is also a large producer of mined rare earths and is the world's major market for rare earths. In 2007, Japan and Southeast Asia accounted for an estimated 24,500 t (37%) of the 66,000 t of contained rare earth oxides (REO) consumed worldwide, with China consuming a further 12,000 t (18%). A decline of these markets could have a serious effect on the supply and demand balance. China supplied more than 60% of the world's rare earth production in 1997. Japan, the largest single consumer of rare earths, is forced to rely heavily on China for its rare earth raw material. Japanese imports of rare earth products from China were valued at a record ¥14.4 billion in 2007 (Roonwal, 2009).

Growing concern over global warming and the potential use of rare earth elements in green technology have made them indispensable. India's ambitious plans for power

TABLE 5.4

Use of REE in High Technology: An Industry-Wise Breakup

End Product	Industry	% Use in 2010
Auto catalyst	Catalytic converters in automobiles	7
Glass additives	Digital camera lens	6
Polishing powder	Polishing material for TV, jewels, silicon chips	14
Magnet	Cellphones, wind turbines	25
Phosphors	LED, computers, and TV display	6
Metallurgy	Nonbattery alloys	9
Fluid cracking catalyst	Petroleum: refining of crude oil	15
Battery alloys	Rechargeable car batteries in hybrid electric automobiles, electronic devices	14
Others		4

Source: Canadian Imperial Bank of Commerce 2012.

production through wind energy may not bear fruit without rare earths. Installation of windmills at a height of about 120 m will need compact motors. Fixing heavy motors at that height can be possible by use of neodymium, iron, and boron. These make the magnets compact and yet powerful (Tables 5.4 and 5.5).

TZML regards the most significant new development in the Indian mineral sand sector to be in Tamil Nadu, where Mineral Deposits Ltd. (MDL) of Australia has outlined proposals to mine substantial deposits in the southeastern part of the state. If this were to go ahead, it would constitute a completely new minerals sand province in India and mark the first direct foreign investment.

The deposits are at Kudiraimozhi, Sattankulam, and Navaladi, located 3035 km southwest of the port of Tuticorin, and are reported to contain a total resource of 100 mt of heavy minerals, including 61 M of ilmenite. According to TZML, two projects are proposed. At Kudiraimozhi, there is an inferred and indicated resource of 370 mt with an average grade of 8.9% heavy minerals. A mining operation producing 270,000 t/y of ilmenite is envisaged, which would provide feedstock for a 135,000 t/y capacity synthetic rutile plant. The total project cost is estimated at US$150 million.

TABLE 5.5

Production of Rare Earth Minerals (t REO Content)

		1996	1997 Update 2011–2012
Monazite	Brazil	200	400
	India	2700	2700
	Sri Lank	120	120
	Total	3020	3220
Bastnaesite	China[a]	55,000	50,000
		20,400	20,000
	Total	75,400	70,000
Xenotime	Malaysia	340	300
Loparite	CIS	6000	6000
	Grand Total	84,800	79,500

[a] Chinese bastnaesite figure includes production of xenotime and monazite. Based on USGS and Mining J., London, Aug. 1998.

The second project is based on the Sattankulam–Navaladi deposits, in which there is an inferred and indicated resources of 610 mt, an average grade of 11% heavy minerals. The project would also include production of synthetic rutile, plus a 50,000 t/y pigment plant with a total investment of the order of US$350 million. TZML says that two Mauritius-based companies have been approved by the Government of India as foreign collaborators in the project Ausind Sands Ltd., and Mauritius Titanium Ltd. (MTL). It is understood that MTL proposes to prepare a bankable feasibility study for each project and that appropriate expertise has been recruited to undertake this work.

Further ahead, Tata Iron and Steel Ltd., the largely privately owned Indian industrial conglomerate, has expressed some interest in entering the mineral sands and titanium metal sectors, but no project has been announced. TZML believes that the entry of Tatas adds considerable weight to India's heavy mineral sand industry.

5.6 Factors Controlling Formation of Beach Placers

Mechanical concentration by natural gravity and separation of heavier minerals from the lighter ones under the action of wave, moving water, and/or wind results in the formation of placer mineral deposits. Placers are classified into (i) eluvial, (ii) fluvial, (iii) aeolian, and (iv) beach placers as per their mode of transportation and site of deposition. Though all modes of placer formations are important, beach placers are significant by virtue of their extent and large tonnage. Some of the geological and geomorphologic factors that control the concentration of heavy minerals along the Indian coast are as follows:

1. *Geological Control:* Types of country rocks, that is, igneous, metamorphic, or sedimentary, and the effect of various geological processes have played a vital role in contributing sediments to form a placer deposit.

2. *Climatic Factor:* The climate of the region has a great role to play in decomposing and disintegrating the rock and mineral fragments that get liberated and concentrated. Tropical to subtropical climate promotes deep chemical weathering along the coastal region. These conditions also favored the formation of laterite, which, in effect, is a process of preconcentration.

3. *Drainage Pattern:* The availability of young and youthful rivers and their high density, coupled with climatic factors, played a dominant role in the supply of material for concentration along favorable locales, especially along the ghat sections of Kerala and Tamil Nadu coasts. The rivers joining the Bay of Bengal on the east coast, however, have attained maturity and in many cases delta systems have developed, for example, Mahanadi, Godavari, Krishna, and Cauvery.

4. *Coastal Processes:* All types of wave velocity, long-shore currents, and wind speed also have their effects in sediment winnowing transport, sorting, and deposition of placer minerals. Emergence and submergence of the coast during geological past also effected the beach placer formation.

Apart from these, numerous other factors that helped in the formation of these deposits are coastal geomorphology, neotectonics, and continental shelf morphology.

5.7 Exploration and Evaluation of Heavy Mineral Placer Deposits

Exploration is a search for unknown, and for placers, the areas of study extend from the slopes in high mountain ranges to the seafloor along the continental shelf. Application of remote sensing technique during the recent past, particularly the satellite and radar images, the infrared photography and multispectra outputs, in conjunction with existing geological maps, has helped in revealing the facts that are generally not apparent by surface observation. A study of aerial photographs further supplements the data. These studies, along with interpretation of available data collected from various sources, will help in narrowing down the target areas for exploration to a reasonable level and could be cost-effective and optimize exploration time (Talpatra, 1999).

Various geophysical methods are being increasingly used in the exploration, in general, and offshore, in particular. The principal ground geophysical methods are seismic, gravity, electrical, magnetic, and radiometric. As a result of mineralogical variations in the deposits, the methods that are applicable to one area may not be suitable for another. Geophysical methods could be used during the reconnaissance survey, and quick interpretation of the data might be helpful in selecting area for detailed investigations.

Exploration is usually undertaken in three stages with each succeeding stage depending on the results of the previous one. They are reconnaissance, scout drilling, close-space drilling, and computation of reserves.

5.7.1 Reconnaissance Survey

Reconnaissance survey helps in covering larger areas in a short span of time and this gives a fairly good information regarding the grade variation along and across the coast/deposit, extent, and inferred reserves in the area. Widely spaced boreholes are drilled by hand-augers down to water table over 500 to 1000 m (along the coast) × 100–200 m (across the coast). This drilling pattern was adopted during reconnaissance surveys for deposits along the east coast of India. A deposit is considered to be economically viable if it can sustain a mining industry for a period of 20 years with an annual production of 1–1.5 lakh tonnes of total heavy minerals. During reconnaissance surveys, field information on various aspects like extent of the deposit, geology of the hinterland, geomorphic controls of heavy mineral concentrations along with infrastructural facilities like road, rail, port, power supply, fresh water resources, and status of land is also collected. Data on climatic conditions are also collected to understand the degree of chemical weathering, which has a direct bearing on the composition of individual heavy minerals and also for selecting mining methods.

In reconnaissance investigations, laboratory work, leading to estimation of slimes, carbonate (shells), and total heavy mineral content, is carried out. Reserves of inferred category are estimated down to a depth of water table. The chemistry of individual heavy minerals, along with their textures and modes, is determined.

5.7.2 Scout Drilling

Scout drilling is undertaken when reconnaissance survey has indicated encouraging results about the deposit in terms of quantity and economic value. The data collected during the reconnaissance would also indicate the course and scope of the investigations to follow. The objectives of the investigation at this stage are as follows:

1. To know the approximate size, depth and grade of the deposit
2. The difficulties that are likely to affect mining and mineral processing
3. Most suitable aspects and spacing of the holes for final grid
4. The most suitable equipment for processing
5. The life of the deposit and consequent milling

In general, the grid lines are normal to the major axis of the mineralized dune, whether it be a frontal or Aeolian dune or an offshore bar. The baseline is marked out, parallel to the trend of the coast and is angled wherever necessary.

5.7.3 Close-space Drilling: Detailed Investigations

Close-space drilling begins for the evaluation of a placer deposit, fairly defined in three dimensions. One of the fundamental difference between preliminary and detailed exploration lies in the density of exploration grid. Normally, the distribution of holes will be extension or development of the original scout drilling. It is gradually closed until the payable zones are delineated. The spacing of holes across the line may be brought down to 30–50 m depending on the width of the deposit so as not to miss the narrow streaks.

In detailed investigations, boreholes are drilled at closer intervals down to bed rock and this gives a three-dimensional picture of the deposit, which helps in mine planning and dredging. The reserves are of measured category.

Sampling: The ultimate purpose of sampling is to obtain a comprehensive set of data with measurements that are reproducible and with observations that may be required for reserves estimation, mine planning, and designing of plant. The samples are collected at every 1.5 m interval up to bed rock and the depth of water table is also marked in each hole.

The reliability of sampling increases with the number of samples taken and the volume of influence assigned to each sample. Hence, a standard set of conditions should be applied to all the samples so as to keep the sampling error within the optimum tolerance limit.

5.7.4 Computation of Mineral Reserves

The tonnage of raw sand in each of the composite sample block is calculated by multiplying the volume of the block by the bulk density of composite sample. From the tonnage of raw sand, the individual heavy mineral reserves are estimated as per the distribution pattern of individual mineral in weight percentages, obtained by the microscopic grain-counting of that particular composite sample. By assuming the reserves of individual blocks, the gross reserve of minerals in the deposit is arrived at. The computation of area and volume in each composite block is of primary importance. In general, there are two methods used for the calculation of area and volume of the composite blocks.

1. *Even Topography:* If the relief of a deposit is less than 1 m, then the deposit is considered as a plain one and in such a case, length × width × av. The thickness of the drilled depth gives the volume in cubic meters. Such a method was adopted for reserve evaluation of the Chavara deposit, where the topographic relief is very low.

2. *Sectional volume method:* This method is adopted when undulated topography is observed across as well as along the coast. The sections are drawn (both longitudinal and transverse) on a graph sheet (on a scale of 1:5000 horizontally and 1:1000 vertically) with the aid of a contour map. The locations of boreholes, their depths, and sampling intervals are plotted on the section. And the sectional area represented by number of small squares is counted. The total number of small squares is multiplied by the scale of the map to obtain the sectional volume. By this way, the sectional volume for all the grid lines representing the composite block is added for a desired depth that represents the composite sample.

5.7.5 Sampling and Drilling Equipments

A number of hand and mechanical drilling equipments are used, and they vary depending on the nature of material and type of coastal morphology. Most commonly used equipments are the hand- and power-driven augers and a type of sludging equipment called Conrod Bunka Drill, made out of aluminum rods, are introduced and are comparatively better than sand drills in terms of progress and depth penetration. Australians are using Reverse Circulatory Drill (RCD), which is fully mechanized and can penetrate harder formations and clay horizons with better core recovery.

5.8 Important Coastal Placer Deposits of India

The principal heavy mineral constituents of the Indian coastal placer deposits are ilmenite, leucoxene, garnet, sillimanite/kyanites, pyriboles, rutile, zircon, monazite, and magnetite. The salient features of only important deposits are presented in the following and for more details about these and other deposits, the reader is referred to the succeeding chapters in this volume that cover some major locations.

5.8.1 Ratnagiri Deposits, Maharashtra

Along the Ratnagiri coast of Maharashtra, beach deposits containing various proportions of limonite and magnetite (Figure 5.3) have been identified. The Deccan basalts are the main source for these placer deposits. These deposits are characterized by finer grain size (1–3 Mt) and 50%–52% TiO_2 content. The beach placer in the Bhatya, Purangad, and Goankhede sectors contain high amount of magnetite (40–60 wt.%) and limonite/hematite (25–35 wt.%) with comparatively lesser concentration of ilmenite (5–25 wt.%) in raw sand. The mode of ilmenite is exceptionally fine and often less than 63 μm. In Kalbadevi and Newre, where the grade of total heavy minerals (THM) is 30%–35%, ilmenite constitutes more than 92% of THM (15%–80% of raw sand), with minor quantities of limonite, hematite, and magnetite. The chemical assay of ilmenite gives 50%–52% TiO_2, 21.2% FeO, 23.9% Fe_2O_3, 0.05% Cr_2O_3, and 0.1% V_2O_5.

5.8.2 Neendakara–Kayamkulam Deposit, Kerala

Occurring between Kayamkulam and Neendakara near Chavara, Kollam district, Kerala, the deposit has an extension of 22 km and width of 225 m. The barrier beach placer is under active exploitation by the Indian Rare Earths Ltd. (IREL). The deposit, explored down to a depth of 7.5 m (average grade 45% T HM), is estimated to contain 2.7 Mt of ilmenite (35% grade), 1.0 Mt of rutile (2.5%), 0.9 Mt of zircon (2.5%), 0.17 Mt of monazite (0.5%), and 2.0 Mt of sillimanite (7%) (Figure 5.4). The model of ilmenite is 177 μm whereas that of zircon and monazite is 105 μm. The garnet content of the Neendakara–Kayamkulam deposit is rather low (0.35% of the raw sand) with mode varying from 250 to 177 μm. Chemically, the ilmenite of Chavara is one of the best with over 60% of contained TiO_2.

5.8.3 Manavalakurichi Deposit, Kanyakumari District, Tamil Nadu

This is one of the oldest known deposits of India, discovered in the early part of twentieth century. Initially, the deposit was mined for monazite content and later for the associated titanium and zirconium minerals as well. Presently, the area is under the mining lease of IREL. The deposit extends for 6 km in length with an average width of 44 m. The total heavy mineral content of this deposit is 39% (up to an average depth of 7.5 m). Ilmenite is the predominant mineral with a grade of 24% (in low-sand) with 1.8% rutile, 0.9% leucoxene, 2% zircon, 1% monazite, 3.5% sillimanite, and 5.5% garnet. Ilmenite from this area contains 55% TiO_2 whereas zircon has 65.5% ZrO_2. The monazite from this

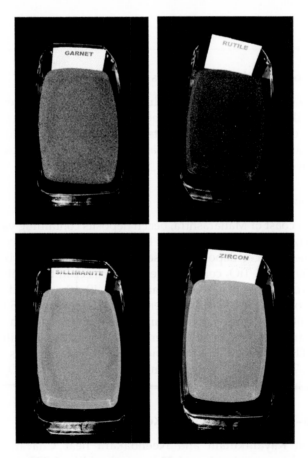

FIGURE 5.4
Purified fraction of heavy mineral sand from the Indian Coast.

FIGURE 5.5
Simultaneous backfilling activity at Chhatrapur Beach. (Courtesy: IRE Ltd.)

area has a total REO content of 58% and 8% ThO_2. Recent studies have indicated that the heavy mineral concentration continues further eastward with a slight depletion in grade. There are also inland placer deposit (Teri sands) in Tamil Nadu. Along the southern coast, two very large deposits of "Teri" sand, and soil (red sediments) occur near Sattankulam and Kudiraimozhi in the southern districts of Tamil Nadu. They are 2–10 m high. These deposits occur in a semiarid and uninhabited rain-shadow region of the Western Ghats, 8–10 km interior from the shoreline. Four such deposits with an aggregate area of 144 sq. km have been identified in parts of the Tirunelveli, Chidambaranar, Ramanathapuram, and Kanyakumari districts of Tamil Nadu (Figures 5.5 and 5.6). The average grade of THM of these deposits is about 10% with a total of 96 Mt and an ilmenite reserve of 58 Mt.

5.8.4 Bhimunipatnam Deposit, Vishakapatnam and Vizianagaram District, Andhra Pradesh

Along the Vishakapatnam and Vizianagaram–Konada sector of the east coast, a sizeable placer deposit has been delineated over a coastal length of 37 km, with a width

FIGURE 5.6
Garnet rich red sands at Sattankulam, Tamilnadu, India.

ranging from 150 to 1000 m (Figure 5.7). The average THM content of this deposit is 19.58% (with ilmenite 9.14%, rutile 0.59%, and monazite 0.09%). The total reserves of heavy minerals in this area are 6.18 Mt with 2.88 Mt of ilmenite. The Bhimunipatnam ilmenite contains 49.1% TiO_2, 18.6% Fe_2O_3, 22.9% FeO, 0.08% Cr_2O_3, 0.59% MnO, 0.4% AI_2O_3, and 0.6% MgO.

FIGURE 5.7
Common example of applications of rare earth's in high technology. (IRE, Chhatarpur.)

5.8.5 Chhatrapur Deposit, Ganjam District, Odisha

Along the Chhatrapur coast, Ganjam district, Odisha, concentration of heavy minerals has been observed for over 18 km length, with an average width of 1.5 km (Figures 5.5, 5.7 through 5.10). The average THM content of this deposit is 20.5% (with ilmenite 8.80%, rutile 0.38%, monazite 0.27%, zircon 0.31%, garnet 6.70%, and sillimanite 3.40%). The total reserves of heavy minerals in this area is 47.25 Mt, with more than 20 Mt of ilmenite. The deposit is under exploration by IREL. The ilmenite of Chhatrapur deposit contains 50% TiO_2, 35% FeO, and 12% Fe_2O_3 (Figure 5.11).

5.9 Heavy Mineral Placer of the East African Coast

Along the African coast, heavy mineral placer deposits occur comprising a variety of minerals including diamond province. They occur on beaches and the adjacent continental shelf zone between 100 and 200 m below sea level. This belt extends to 450 km on the south and 300 km north of Orange River. Along the Namibian coast and South Africa, the placer deposits comprising an assemblage of titanium, thorium, rare earth elements, and zirconium occur. These placers are being mined for several years. Placer deposits occur also on coast of Madagascar. It is estimated that this rich placer belt contains more than 70 million tonnes of ilmenite at an estimated average ilmenite grade of 4.14%, and the Moma beach contain almost 60 million tonnes of ilmenite along onshore and near shore the coast of Mozambique.

FIGURE 5.8
Dredge and West Upgradation Plant on Chhatrapur/Gopalpur Coast at Chhatrapur Beach. (Courtesy: IRE Ltd.)

FIGURE 5.9
Dredging operation at Chhatrapur Beach. (Courtesy: IRE Ltd.)

FIGURE 5.10
Showing different layers of heavy minerals in the beach sands in west coast of India. (Courtesy: P.K. Mallik.)

FIGURE 5.11
Sillimanite rich beach placer in Chhaterpur, Gopalpur, Odisha, India.

Placer deposits in the Indian Ocean have been worked since ancient times, at locations such as Richards Bay, South Africa, and the Kerala state in India. Tin placers have been collected by offshore dredges. One such large offshore potential deposit of ilmenite lies off the mounts of the Zambezi river in Mozambique. Extensive bottom sampling, vibrocoring as well as geophysical measurements led to discovery of our large heavy mineral sand deposits containing 50 million tonnes of ilmenite, 1 million tonnes of rutile, and 4 million tonnes of zircon in a water depth upto 60 m off Zambezi delta. These sands were possibly deposited during the Holocene (Beirsdorf et al., 1980).

5.10 Tin Placers of Southeast Asia

A widespread suite of tin placers occurs on the continental margins of Asia such as tin placers in the Southeast Asia offshore. Of these various deposits, placers of the tin mineral cassiterite offshore Southeast Asia are the principal deposits that are being mined for several decades. There source is derived from the weathering of granitic rocks of Myanmar, Thailand, Malaysia, Laos, and Cambodia. Besides the tin deposits, there are several placer deposits of light heavy minerals (ilmenite, rutile, magnetite, zircon, garnet, and monazite) that are present on beaches and offshore of the Indian subcontinent (Rajamanickam, 2000; Roonwal, 1986) and People's Republic of China.

Tin placer offshore dredging in South Asia is one of the largest marine metal mining operations in the world and has been active for several years. The large-scale tin placer mining is carried out in Indonesia and Malaysia. These submerged tin placers are submerged and buried fluviatile and alluvial fan deposits from around Bangka Island, the east coast of Belitung Island, and off Karimun and Kundur Islands near Singapore. The mining is done by ladder dredges ranging in bucket capacity from 7–30 cubic feet. The maximum dredging depth is currently 50 meters below sea level (BSL) using the largest dredges.

In Indonesian offshore placer, tin deposits are explored to maximum dredging depth to 90 m BSL; alluvial tin resources are known to exist below 80 m depth BSL. Even if new deep dredge is commissioned, it will only be able to dredge to 70 m depth. Consideration needs to be given in developing conceptual designs for commercial dredgers at even greater depths. These important tin placer discoveries of fluviatile tin placers continue to be made, particularly off the west coasts of Bangka and Kundur Islands and the east coast of Belitung Island. Some of these have offshore primary mineralization sources associated with submerged granites, at a limited dredging depth BSL. The most appropriate digging technology, which is technically feasible beyond 50 m depth, uses a reinforced articulated ladder, attached midway on the dredge pontoon, comprising two hinged sections, with a bucket wheel cutter suction excavator operating at the cutting face. Computer modeling showed that it is best to use hydraulic transport for lifting the tin ore from the seafloor to the surface and to use barges to transport the tin ore to land-based processing plants. Double-suction grab dredges and trailing suction hopper dredges had also been considered but were rejected as being less suitable.

5.11 Resource Position

Among the important world shoreline placer mineral deposits, mention may be made of the deposits in the east and west coast of Australia, southern part of China, Sri Lanka, Malaysia, Madagascar, Vietnam, Egypt, Sierra Leone, Republic of South Africa, Guinea, Ivory Coast, Brazil, southeastern United States, and Canada. Most of the deposits of Australia, the United States, South Africa, and China are of low grade (4.5% THM), However, the extent of these reserves and the innovative advances in exploration techniques being adopted in some of these developed countries have made them leading nations in exploitation and export of these mineral commodities. Ilmenite in Canada is mostly mined from the hard rock.

The mineral sand deposits of India are a true reflection of their varied provenance ranging from Archaean to Recent. Favorable geological, geomorphological, climatic, and other factors have led to the formation of these deposits that are probably among the richest and largest of their kind. Tile mineral assemblages vary widely from near monomineralic as near Ratnagiri to multimineral-bearing as in other areas. They vary in their grade, mineral assemblage, chemical composition, and texture, all reflecting their diverse source and provenance. The source for heavy minerals for the placer deposits of India are khonadalites, charnockites, Deccan traps, granites, and gneisses. The reworked Tertiary sandstones as at Warkala, Cuddalore, and Rajamundry also contribute to these placer deposits.

India has last resources and reserves of heavy mineral sand; this natural wealth is not being fully utilized. India was one of the major suppliers of ilmenite, rutile, and zircon up to 1940s. The political upheavals of the World War II and later the policy of the government to regulate and conserve mineral resources led to a slow and steady downfall of this flourishing industry. Another factor that is responsible for such underutilization is the discovery of new deposits in other parts of the world, particularly in Australia, South Africa, the United States, and Canada. However, these countries have more or less exhausted all their rich deposits. Ilmenite containing higher TiO_2 (60%) content, which is the mainstay for the sulfate process, is being exhausted very fast. However, the strict environmental control has given way for chloride process of pigment production, which uses upgraded ilmenite, either in the form of synthetic rutile (92% TiO_2) or titanium slag (80%–85% TiO_2) as its feedstock. For both these processes, ilmenite of 45%–50% TiO_2 is the raw material. The working grade of mineral sands world over has also registered a decrease and is of the order of 4%–5% of total heavy mineral concentration, with ilmenite of 50%–52% TiO_2 content.

India has reserves of ilmenite of this grade all along its eastern and western seaboard. The only deposit where TiO content of over 60% is at Chavara, with lower contents of Cr_2O_3, V_2O_5, PO, CaO, U, and Th. The world scenario of reserves and production indicates that although India stands among the top few countries in reserves, it is among the last few in production. Even smaller countries are doing better with regard to mining and export of these minerals. The demand for titanium pigment, a major consumer of ilmenite, is likely to go up by 25% in the next 10 years. Likewise, the demand for garnet is also on increase as a replacement for silica in sand-blasting and for other uses. The present time is, therefore, most appropriate for India to enter the world market (Table 5.6).

TABLE 5.6

Operational Marine Metallic Heavy Mineral Deposits

Name	Commodity	Type of Deposit	Water Depth (m)	Location Latitude, Longitude
Heinze Basin	Tin, tungsten	Placer	16–30	Myanmar, 14.7°N, 97.8°E
Richard's Bay	Titanium, zirconium	Placer	0–30	South Africa, 28.8°S, 32.0°E
Fort Dauphin	Titanium, thorium, rare earths, Zirconium	Placer	0	Madagascar, 25.0°S, 47.0°E
Kanyakumari Manavalakurichi	Titanium, Zirconium, thorium	Placer	0	India, 8.2°N, 78.5°E
Chhatrapur	Titanium, zirconium, thorium	Placer	0	India, 19.4°N, 85.0°E

Source: Modified from Lenoble, J. P. et al., 1995. *Marine Mineral Occurrences and Deposits of the Exclusive Economic Zones*, MARMIN: A Data Base. Editions IFREMER, p. 274; Rona, P. A., 2008. *Ore Geol. Rev.*, 33: 618–666. Doi:10.1016/j.oregeorev.2007.03.006.

Yet another and more profitable way of exploiting these minerals is by setting up down-stream industries for these minerals instead of exporting in raw form. This involves, apart from mining and milling processes, establishment of synthetic rutile or titanium-slag plant, and the manufacture of titanium pigment or metal. Today we have a healthy and conducive industrial environment than ever before. With the recent new mining and industrial policies, it will not be long before India takes its rightful place in tile mineral sand market of the world.

References

Ali, M. A., Krishnan, S., and Banerjee, D. C., 2001. Beach and Inland heavy mineral sand investigation and deposits in India—An overview. In: *Special Issue on Beach and Inland Heavy Mineral Sand Deposit of India*. Directorate of Exploration and Research for Atomic Mineral, Hyderabad, Vol. 13, pp. 1–21, 170.

Ali, M. A., Vishwanathan, G., and Krishnan, S., 1998. Beach and inland Sand Placer deposits of India. In: *National Symposium on Later Quaternary Geology and Sea Level Changes*, Cochin, Directorate of Exploration and Research for Atomic Minerals, Extended abstract, November 4–6.

Atomic Minerals Directorate, 2001. Dept. of Atomic Energy. Govt. of India spl. vol. on "Beach and inland heavy minerals and deposits of India," p. 116.

Baturin, G. N., 1982. *Phosphorites on the Seafloor: Origin, Composition and Distribution*. Elsevier, Amsterdam, p. 28.

Beirsdorf, H., Kudrass, H. R., and von Stacklberg, U., 1980. Placer deposits of ilmenite and zircon on the Zambezi shelf. *Geol. Jahrb.*, D-36: 5–85.

Canadian Imperial Bank of Commerce 2012. Reports.

Emery, K. O. and Noakes, L. C., 1968. Economic placer deposits of the continental shelf. *ECAFE Tech. Bull.*, 10: 95-III.

Hein, J. D., 2008. In: Mineral Deposits of the Sea. *2nd Report of ECOR Panel on Marine Mining*, p. 39.

Kudrass, H. R., 2000. Marine placer deposits and sea-level changes. In: D. S. Cronan (ed.). *Handbook of Marine Mineral Deposits*. CRC Press, Boca Raton, FL, pp. 3–26.

Lenoble, J. P., Augris, C., Cambon, R., and Saget, P., 1995. Marine mineral occurrences and deposits of the exclusive economic zones, MARMIN: A data base. Editions IFREMER, 274 pp.

Mahadevan, C. and Sriramdas, A., 1948. Ilmenite in the beach sands of Visakapatnam Dist. *Proc. Indian Sci. Congr.*, 27A: 275–278.

Mallik, T. K., 2007. Mineral resources of the beaches and shallow sea annual Indian coasts. *Indian Geol. Congr. Bull.*, I: 82–87.

Rajamanickam, G. V., 1996. Coastal placer development and environmental management. In: S. Z. Qasim and G. S. Roonwal (eds.). *India's Exclusive Zone*. Omega, New Delhi, pp. 65–73, 238.

Rajamanickam, G. V. (ed.). 2000. *Handbook of Placer Mineral Deposits*. New Academic Press, Kolkata, p. 327.

Rao, B. L. and Reddy, D. R. S., 1993. *Lime mud occurrence off North Andhra Pradesh coast. Mar. Wing GSI News Lett.*, 9(2): 17.

Rao, B. R., Mohapatra, G. P., Vaz, G. G., Reddy, D. R. S., Hari Prasad, M., Mishra, U. S., Raju, D. C. L., and Shankar, J., 1992. Inner shelf placer sands off north Andhra Pradesh. *Mar. Wing GSI News Lett.*, 8(1): 11–12.

Rao, P. G., 1988. Sediments of the near shore region off Needkara, Kayankulam coast and Ashtamudi and Vatala estuaries, Kerala, India. *Bull. Natl. Inst. India*, 38: 513–551.

Rona, P. A., 2008. The changing vision of marine minerals. *Ore Geol. Rev.*, 33: 618–666. Doi:10.1016/j.oregeorev.2007.03.006.

Roonwal, G. S., 1986. *The Indian Ocean: Exploitable Mineral and Petroleum Resources*. Springer-Verlag, Heidelberg, Vol. XVI, p. 198.

Roonwal, G. S., 2009. Beach sand resource and industry in India. In: K. L. Srivastava (ed.). *Economic Mineralization*, Scientific Publishing House, Jodhpur, pp. 248–254, 390.

Siddique, H. N. and Mallik, T. K., 1973. "The Investigation of Concaveness Sand Deposits in the Lagoons of Laccadive." Unpublished Rept 481.

Talpatra, A. K., 1999. Heavy mineral placer deposits of India with special reference to 3D modeling in parts of offshore areas for resource evaluation: a suggestion. *Indian J. Geol.*, 71: 105–115.

TZ Mineral International Pty Ltd. (TZMI), 2000. *Mineral Sand Deposits*. West Perth, West Port, Australia, p. 9.

Vaz, G. G., Mishra, U. S., Biswas, N. R., Vijay Kumar, P., Krishna Rao, J. V., Sankar, J., and Faruque, B. M., 1993. Lime mud continental shelf edge and slope off Kachch. *Indian J. Mar. Sci.*, 22: 209–215.

6

Shallow Oceanic Nonliving Resource: Petroleum and Hydrocarbon

6.1 General

Petroleum is so important an energy source that it is appropriately called "black gold" to signify its place in the present-day life and industrial growth. It is hard to believe that how style of life would be without energy. Today, we generate electricity through natural gas, and diesel and petroleum. All our transport by air, sea, and land is dependent upon petroleum or gas form of hydrocarbon. In the urban and rural life, and in small-scale industrial units, liquified petroleum gas (LPG) is used for cooking and heating purposes.

The energy need in the country is witnessing a quantum jump, so also is the generation of power. The hydrocarbon production has increased many folds in the last few decades. But the consumption and demand has increased even faster. At present, we have a big shortage of production and much of the needs are met by import of hydrocarbons from the Gulf region and other available sources. Indeed, a small reaction to our dependency on energy can be felt during the summer months in northern India, when the temperatures go up to 40°C and higher, the consumption of energy increases far more than its production. This leads to frequent tripping and load shedding. The last 30 years has seen enormous growth of the extraction of oil gas from below the continental shelf. In 1986, 74% of the crude oil production came from offshore.

Hydrocarbons today play a critical role in sustaining industrial and therefore economic growth. Their production and consumption have rapidly increased in the last decade. There is reason to believe an equal increase in demand considering the all-round growth in the automobile industry, with the permission being given to industrial units to install and generate their own power units to keep pace with the overall growth of the industry. The present production of hydrocarbons in India is about 35 million tonnes of oil and 25 billion cubic meters of gas. The national consumption of oil is 65 million tonnes.

This amount is very significantly large because it constitutes nearly 3% of gross national product (GNP) and accounts for nearly 30% of the total national imports. Consider the trade as such; this huge bill accounts for exports to the extent of 35%. This difficult situation has to be faced by India; one may as well remember that at present the per capita consumption of petroleum products in India is very low, about one-eighth of the world average. As mentioned above, this is rapidly changing because of new liberal policy of industrial growth. Thus, the demand for hydrocarbons in the coming decades will certainly increase by several folds.

India's demand for fossil fuel by 2040 is estimated to grow at a compounded annual average growth rate (CAAGR) of 3.6% for oil and 4.6% for natural gas to keep pace with the projected economic growth. The oil import dependency of the India is predicted to rise to 90% from the current 70%. To meet the increasing energy demand, the government of India has adopted several policies such as opening up of upstream exploration through competitive bidding under NELP during 1997–1998, marginal field policy during 2015, and is now moving toward open acreage policy model for future acreages under hydrocarbon exploration licence policy (HELP) (Table 6.1).

Hydrocarbons are generated by the decay of organic matter in sedimentary strata when it is subjected to overburden weight and pressure to generate a high temperature sufficient to mature it. Hydrocarbons thus generated are retained in situ when capping strata of impervious rocks and suitable structural traps exist. Since the continental rise of which the continental slopes in the deep ocean are composed of erosion products of the continents contain good amount of organic material. The thickness of sediments in the continental shelf rise is often several kilometers and in some places such as the Indus fan in the Arabian Sea and the Bengal fan in the Bay of Bengal, the erosion products of the Himalayas accumulate. It is a very thick pile of sediments of the range of several thousand meters (Dwivedi 2016). This makes it attractive for high possibility of finding hydrocarbons (Figure 6.1).

TABLE 6.1

Classification of Indian Offshore Sedimentary Basins

Category	Type	Basins	Status
II. Rifted Setting	II Marginal Aulacogen	1. Cambay Basin	Oil and gas producing basin
	II Pericratonic Rift	1. Kutch Basin	Oil and gas producing basin
		2. Saurashtra Shelf	Oil and gas indications
		3. Surat Basin	
		4. Bombay–Ratnagiri Basin	Oil and gas producing basin
		5. Kondan Basin	
		6. Kerala Basin	
		7. Western Bengal Basin	
		8. Mahanadi Basin	Gas discoveries—Emerging basin
		9. Krishna–Godavari Basin	
		10. Palar Basin	Oil and gas producing basin
		11. Cauvery Basin	
		12. Mannar Basin	Oil and gas producing basin
	II Intra-Cratonic RM	1. Pranhita–Godavari Basin	–
		2. Mahanadi Graben	
		3. Purnea Graben	
	II Miogeoclinal Prism	1. Deep Water Basins of West coast	–
		2. Deep water Basins of East coast	
		3. Andaman Basin	
	III Remnant Ocean Basin	1. East Bengal Basin	Gas producing basin Gas indications

Source: After Biswas, S. K., Bhasin, A. L., and Ram, J., 1993. Classification of Indian sedimentary Basins in the framework of plate tectonics. *Proceedings Second Seminar on Petroliferous Basins of India*, Indian Petroleum Pub. Dehradun, 1: pp. 1–46.

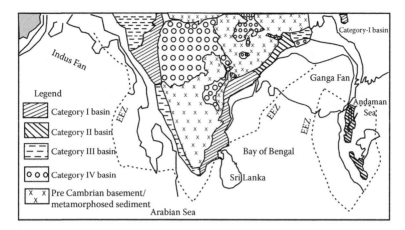

FIGURE 6.1
Showing off-shore sedimentary basins in the Arabian Sea and Bay of Bengal. (After DGH, 2015. Hydrocarbon exploration and production activities, India, 2014–15. pp. 14–17, 100–108, 112–118, 152–153.)

6.2 Potential Hydrocarbon Basins in the West Coast of India: Arabian Sea

The basins in the western continental margin of India (WCMI) situated on the western passive margin of the Indian plate evolved during the separation of Madagascar–Seychelles in Late Cretaceous between 90 and 65 Ma. The rifting started in Late Cretaceous with syntectonic Deccan volcanic activity, which continued till Early Paleocene. The Early Paleocene to Early Eocene hiatus was the period of the rift–drift transition marked by a widespread unconformity in all the basins. The postrift thermal cooling resulted in sagging and the basins evolved into a marginal sag initiating marine transgression (Figure 6.2).

Coast parallel ridge-depression couplets, Kori-Comorin Ridge/Depression and Laxmi–Laccadive Ridge/Depression, crossed by first-order transverse basement arches are the major features of the WCMI structure (Biswas, 1989). The transverse arches, from north to south, are Saurashtra, Bombay, Vengurla, and Tellicheri arches, which divide the shelf into five offshore sub-basins—Kutch, Surat, Ratnagiri (Bombay offshore basin), Konkan, and Kerala basins (Biswas and Singh, 1988). Each sub-basin extends offshore across the shelf, depressions, and ridges. Thus, a shallow shell horst-graben setting, a shelf margin depression and a ridge, a deep slope-parallel depression, and an outer ridge (Laxmi) separating the basin and abyssal plain are the characteristic structural domains. Kerala, Konkan, and Ratnagiri basins are confined to offshore. The West Coast Fault terminates these basins along the coastline. The Kerala–Konkan Basin, though well explored, has no commercial accumulation that could be located till date. Awaiting discovery, this basin is listed as a Category III basin.

The Mumbai Offshore Basin is the most prolific of all the hydrocarbon-bearing provinces in India. Bounded by Diu and Narmada Faults and Deccan Trap outcrops to its north and east, the pericratonic Mumbai Offshore Basin extends toward west parallel to the western continental margin of India up to the Western Margin Basement Arch. The northeast–southwest trending Vengurla Arch separates the basin from the Kerala–Konkan Basin on its south. On the basis of structural elements and the nature of sediment fill, which influences characteristic hydrocarbon generation and entrapment patterns in different sectors, the basin is subdivided into a number of blocks, viz. Tapti–Daman block, Diu block, Heera–Panna–Bassein block, Mumbai High-Deep Continental Shelf block, Shelf Margin block, and the Ratnagiri block.

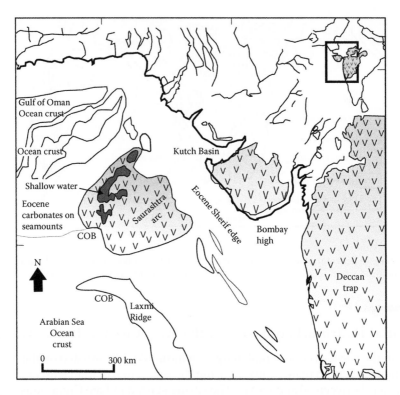

FIGURE 6.2
Tectonic map of west coast of India showing the Kutch Basin. (From different sources.)

Based on tectonic-sedimentary history and alignment of structural elements, the basin is subdivided into six major units, three of which are characterized by paleo-platform set up, viz. the Mumbai High-DCS platform, the Heera–Panna–Bassein Platform, and the Ratnagiri platform, which were dominantly filled up by carbonate sediments. Three of the other units, viz. Murud low, Saurashtra rise–Diu depression, and Surat depression, are mostly characterized by clastic sedimentation acting as the dominant hydrocarbon generation centers for the basin. The Tapti–Daman block basin is developed in the northern part of Bombay Offshore Basin in front of a narrow Cambay Gulf. This has a sedimentation history from Paleocene time onward. Middle Eocene–Oligocene deltaic sediments of proto-Narmada in Cambay Basin and Late Oligocene–Miocene sediments in Tapti–Daman sector deposited as a result of progradation of the delta through the narrow Cambay Gulf.

The major hydrocarbon fields of the basin include Bombay High, Bassein, Panna, Mukta, Neelam, Heera, South Heera, Ratna, D-1 fields, Tapti (north, middle, and south), and C22/24 structures. Hydrocarbon accumulations in BH–DCS–HPB (Bombay High–Deep Continental Shelf–Heera-Panna-Bassein) sector are structurally controlled, and in general occur in carbonate reservoirs ranging in age from Middle Eocene to Middle Miocene with a few exceptions of stratigraphic combination plays in clastic reservoirs. While major play zones occur in Middle Eocene and Early Miocene plays, hydrocarbon zones have also been identified in Paleocene to Oligocene plays whereas in Tapti–Daman sector, Daman (Late Oligocene) and Mahuva (Early Oligocene) plays are well established. Targeting new objectives in established areas and looking for extension of known plays besides exploring deeper plays has always been the essence of exploratory activities in Mumbai Offshore Basin.

Recent successes in basement and basal clastic section from a number of wells on the western and southern periphery of Mumbai High have seen renewed focus and exploration initiatives being adopted for exploration of this play. With integrated interpretation of seismic and well data including special analysis of cores and image logs along with application of state-of-the-art technologies for fracture imaging, a significant improvement has been made in the identification and mapping of probable fracture zones in basement.

The Kutch Basin located on the western continental margin covers an area over 80,000 sq. km both onland and off-shore. The basin has a setting of rift and passive margin basin, which has evolved through multiple rift phases after the separation of the Indian and African plates. Exploration commenced way back in 1972 and a decade later, oil was discovered in the offshore from Eocene Limestone. The onland part of the Kutch Basin appears to be an arm of a failed rift system and is restricted by the Allabandh fault toward the north and extends up to the North Kathiawar fault toward south. The structural trend on the onland part of Kutch Basin essentially trends in an east–west direction whereas in the offshore part of the basin, the dominant structural trend appears in a northwest–southeast direction. This indicates that probably the separation of the Indian and African plates has a more significant imprint in the offshore area. Based on the study of recent gravity magnetic data, it appears that the continental oceanic boundary took an easterly swing around Kutch Basin, distinctly indicating the possible presence of Mesozoic sediments below the Deccan traps even in the deep-water areas. The most important plays in the Kutch Basin are sub-trappean Mesozoics and Early Tertiary sediments.

Remnant ocean Basin of Bengal (Biswas et al., 1993), Mahanadi, Krishna–Godavari (KG), Palar, Cauvery, Mannar rift Basins, and Andaman. Of these, KG and Cauvery (including Palar and Mannar Basins) are category I basins and Mahanadi and Andaman are category II basis. The Bengal Basin still remains a category III basin.

6.3 Hydrocarbon Potential along the Eastern Offshore in the Bay of Bengal

The Eastern Coast Basins define the present coastal and offshore structural framework that evolved since Late Jurassic related to the breakup and drifting of Indian and Antarctic plates (Biswas, 2008). It appears that the eastern continental margin evolved by segmentation into a rifted northern part as well as southern sheared part (Shyam Chand and Subrahmanyam, 2001). However, the Krishna Godavari rifted part is different, being characterized by sedimentary basins having coast parallel horst-graben structures. The Cauvery Basin along the sheared segment shows development of pull-apart basins and intervening ridges oblique to the north–south shoreline. The basins formed on the eastern passive margin of the Indian plate in a divergent setup. The continental shelf is narrow here and the basins pass into deep-water slope to abyssal Eastern Ocean Basin (EOB) of Bay Bengal (400–2000 m) within a short distance (Figure 6.3). The basins from north to south are the following.

Along the east coast, in the Andhra Pradesh coastline, the Krishna–Godavari Basin constitutes a typical passive margin basin. This basin has a dual-rift province evolution history. The basin comprises a wide array of sedimentary facies from early Permian through Cenozoic with the analogous outcrops defining the basin limitation, along the

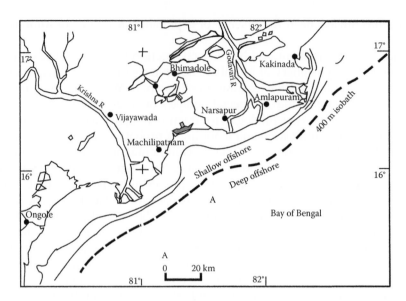

FIGURE 6.3
Offshore hydrocarbon potential along Krishna–Godavari Basin. (Compiled from different sources.)

northwestern part of the basin. It is spread over an area of 28,000 sq. km. in the onland and 145,00 sq. km (24,000 sq. km. up to 200 m isobath) in the Bay of Bengal offshore. The prospective area for oil and gas exploration in onland covers the three coastal districts of East Godavari, West Godavari, and Krishna and in Offshore till 85°E Ridge. Efforts by both NOCs and private oil companies have unlocked huge hydrocarbon reserves.

The basin is estimated to have a sediment thickness in excess of 7.0 km. It can be concluded that this basin comprises linear horst-graben system, growth fault/rollover, and block tilting along synthetic fault over intrashelf regime followed by toe thrusting, exhibiting typical evolution from the rift basin to the passive margin basin. The habitat of hydrocarbons spreads over a wide stratigraphic spectrum from Triassic to Pliocene and geographical distribution, onland, and offshore including ultradeep water domain (Figure 6.4).

FIGURE 6.4
Offshore platform operating in the Arabian Sea. (Courtesy: ONGC of India.)

The recent discovery of oil from synrift play in the Krishna river mouth area (Nizampatnam bay) has opened up a new area, albeit with a challenge of tight reservoirs. The basin has the distinction of reporting maximum number of discoveries in the last decade (DGH: 2005–2015). Subsequent to the discovery of a large Dhirubai gas field (DGH, 2015) by Reliance Industries Ltd., ONGC notified a large gas field from the ultradeep waters in 2006. As such, this basin has shown a very high degree of hydrocarbon potential, particularly in deep waters off the Godavari river mouth, essentially from Mio-Pliocene and Pleistocene Formations. Similarly, recent discoveries of gas from Machilipatnam bay area from Eocene–Pliocene formations from shallow water segment endorses for sustained exploration. Based on the trend of discoveries, there are good chances of discovering large fields, especially from deep-water segment. In KG, deep waters, mainly slope–channel–levee complex, debris flows, low-stand wedge, and basin floor fan complexes remain the major targets. In shallow water-growth fault-related structures, channel fills combination traps and upper slope fans still remain as attractive plays, particularly in the delta-slope transition. Onland, the deeper synrift plays remain the major attractive play. Based on the emerging exploration trend, sands within the Lower to Upper Cretaceous Raghavapuram play attain importance for further exploration.

Cauvery Basin is a pericratonic basin, evolved due to rifting between India and Sri Lanka during the breakup of Eastern Gondwana land during Late Jurassic–Early Cretaceous and subsequent drifting (Late Aptian) of Indian plate from Gondwana land along northeast-southwest oriented Eastern Ghat trend. The rifting has created several horsts and grabens. The present-day horst and graben picture of Mesozoic–Cenozoic stratigraphic column have been related to two principal tectonic episodes, namely extension stage during Late Jurassic to Early Cretaceous and thermal subsidence stage during late Cretaceous to Cenozoic.

Mahanadi sedimentary basin is located in the Eastern Passive Continental Margin of the Indian subcontinent. Geographically, it is flanked by the Bengal Basin in the northeast and Krishna–Godavari Basin in the southwest. The onland part of the basin encompasses the deltaic plains of Mahanadi river and its distributaries lying between Jagannathpur in the northeast and Chilka Lake in the southwest, which extends to offshore Bay of Bengal, covering a total area of about 2,60,000 sq. km. Tectonically, Mahanadi Basin is a polyhistory basin starting as a northwest–southeast trending Permo-Triassic Gondwana rift characterized by a failed arm superposed by Late Jurassic–Early Cretaceous northeast-southwest trending, east coast rifting, and finally leading to passive margin basin set up during Late Cretaceous to Recent period. Various subbasins/depressions with northeast–southwest trend are present, which include Cuttack–Chandbali depression, Paradip depression, Puri depression, Northern and Southern offshore depression with sediment thickness varying from 5000 m onland to more than 10,000 m in the offshore basin.

6.4 Hydrocarbon Potential in the Andaman Sea

India's energy sources are at present limited because they have remained unexplored. The energy produced is also not gainfully used because of loss in transmission and other political factors such as overlooking unauthorized withdrawal of electricity by slum people. The hydrocarbon exploration requires modern technology for exploration. This is receiving attention of ONGC and DGH. Added efforts are needed to locate hydrocarbon deposits to meet the increasing need of hydrocarbons. The major offshore discovery in the Bombay High region is already now "old." Exploration efforts in the immediate offshore basins on the west (Cambay, Lakshadweep) as well as others on the east coast have given potential results. Details are given in Roonwal (2015).

The Andaman Sea Basin lies between oil-bearing Indonesian basin to the south and Myanmar basins to the north. It has been explored by many wells, resulting in the discovery of the accumulation of noncommercial gas. The Paleogene sequence has generally remained unexplored. Hydrocarbon potential in the Andaman Sea appears to be confined to the fore-arc basin. The basin has a 4000–6000 m thick succession of Cretaceous to Quaternary marine sediments, overlying the Cretaceous ophiolites basement. Extrapolation of surface geological mapping on the Andaman group of islands, as well as the data from drilled wells in this fore-arc offshore give a variation in the stratigraphy and lithologs between the two areas. Within the fore-arc, rapid change in the stratigraphy in short distances are found and has suggested a presence of major Neogene–pre-Neogene unconformity as well as a general paucity of reservoirs in Neogene succession. Major information on stratigraphy has been obtained through multichannel seismic data. This has shown that the Paleogene and older sections are highly distributed, with poor, chaotic, natural, and isolated broken reflections, which are in sharp contrast to the fair reflection obtained in relatively less distributed Neogene sediments. Exploratory wells drilled by the ONGC of India in the fore-arc have shown gas in Middle Miocene limestone, which is type III Kerogene and support hydrocarbon generation in the basin. Further attempts are needed for the Paleogene and other sedimentary sections where the presence of good reservoirs and mature organic matters are established. Geology and tectonic of the Andaman Sea is given in Chapter 4.

The area of Andaman Nicobar Island is approximately 6000 km^2. The offshore extension forms a sedimentary basin, which occupies an area of more than 75,000 sq. km including the deep-water segment. The basin formed in intraoceanic setting as an island arc in a convergent setup where the Indian plate is subducting under the Burma–Malayan plate since Cretaceous time. It is a typical island arc with an outer trench, fore-arc basin, an inner volcanic arc, and a back-arc basin. The Andaman fore-arc basin is sitting on the subduction prism represented by outer sedimentary arc (Roy and Das Sharma, 1993). Thick Paleogene flysch overlain by shallow marine Neogene sediments are exposed in the islands of Andaman and Nicobar.

The first gas discovery from Miocene Limestone (AN-1-1) in the islands and oil/gas discovery in the mud volcanoes indicate the presence of hydrocarbons in the Tertiary sediments. Two petroleum systems are envisaged to be operative in the area: (i) the biogenic shallow gas and (ii) the deep thermogenic system. These petroleum systems are envisaged to be operative in fore-arc basin where exploration wells have been in Indonesia and Yadana and Yetagun fields of Myanmar. The main source rocks in this basin are the Cretaceous and Paleocene shales of Baratang Formation. The reservoir rock established so far is the Mio-Pliocene Long Formation. The presence of a completely unexplored rift setup in the southern part of eastern extension of Andaman Island is the future exploration target.

6.5 Gas Hydrates Resources in the Bay of Bengal

Gas hydrates in the present time could be an attractive, nonconventional source of the gaseous hydrocarbons. Gas hydrates generally occur in deep oceanic sediments in tropic regions in disseminated, dispersed, nodular layered, or massive forms as the recent discovery in the Bay of Bengal suggests. The occurrence is controlled by conditions of temperature, pressure, and the composition of the available gases. Information suggests that, in the Andaman Sea, a complex series of Neogene fold packages the progressively underthrusting from east to west in a north–south trend. It appears that the area around South Andaman in the fore-arc region was subjected to maximum subsidence in the Neogene period. These Neogene sediments contain good organic matter but are immature, suggesting biogenic gas prospects within the younger rock (Chandra et al., 1998; Roy and Sharma 1993).

ONGC of India has plans to conduct deep-water resource mapping for resource evaluation. The Indian Oil Ministry is planning to coordinate for the gas-hydrate resource mapping exercise by drilling for cores at locations in the deep-water of the Andaman Islands and Andhra Pradesh. The Directorate General of Hydrocarbons of India has plans for an expression of interest for the deep-water drill ship that can drill in the depth of 1000 to 2000 m. Shallow and high-resolution seismic surveys are being conducted prior to drilling.

Methane gas hydrate in the seabed shallow sediments is a prolific source of gas containing both biogenic and thermogenic gas. In India, the presence of gas hydrate has been established in deep offshore of KG, Mahanadi, and Andaman areas (DGH, 2010). National Gas Hydrate Program of the government of India conducted the first expedition in 2006 and the second expedition of gas hydrates commenced in 2015. Intensive R&D studies are being carried out to evolve technologies to produce gas from frozen seabed sediments and India is ranked third in this endeavor after the United States and Japan.

The total areas of sedimentary basins in India span about 2120,000 km^2. Out of this about 1400,000 km^2 is onshore, whereas 320,000 km^2 is offshore in the water depth upto 200 m and 400,000 km^2 in the water depth more than 200 m. Figure 6.1 shows the major sedimentary basins, which have been found to be of commercial interest for hydrocarbon deposits. It is said that most of the basins have a good sediment thickness of around 5000 m and also with abundant marine argillaceous facies. They have been classified and marked through bore-log and geological consideration, though one may be somewhat careful to note that all of them do not represent a sedimentary basin in true sense.

Hydrocarbons and its products are important for the industrial and overall growth of any nation in the modern times. Hydrocarbon deposits are located in overall offshore and onland location world over. In India, hydrocarbon deposit was first located in Assam region and since 1947 efforts onland, oil in offshore areas have been located. In the offshore areas, Bombay High has been a success. Efforts to survey several geologically potential areas along the west and east coasts have been in progress for some years now. While several moderate-size gas deposits have been located, no major petroleum field has been found. One must keep in mind that the exploitation for hydrocarbons is a very high risk and capital-intensive undertaking. In the offshore areas, the technology is even far more riskier. The availability offshore platform is always a constraint. Even if a deposit is discovered, its development to exploitation level on a commercial basis is equally highly capital intensive and demands high-risk precision work.

6.6 Significance of the Andaman Sea

It is important to give attention to the Andaman Sea because of its metallic minerals and hydrocarbon resources (Roonwal, 2015). The Andaman Sea in the northeast Indian Ocean, lies between 6°–14°N and 91°–94°E is heavily sediment piled (upto 6000 m), actively extending back-arc basin on the Indo-China continental margin. Therefore, the Andaman Sea represents an extensional basin. The opening of the Andaman Sea began 13 my ago (Middle Miocene) at a rate of about 37 mm/yr (Curray et al., 1982), with total opening of the basin has been about 460 km. It now occupies about 800,000 km^2 (Rodolfo, 1969). The bathymetry of the basin is complex. To the west lies the Andaman–Nicobar Ridge, which marks the boundary between the Indian and Eurasian–China plates, and to the east the Malay continental margin. The main topographic features of the Andaman Basin comprise (i) the Central Andaman Rift, (ii) the Central Andaman Trough, (iii) the Deep Trough, and (iv) Barren Seamount Complex, Alcock Seamount, and Sewell Seamount (Rodolfo, 1969). In the Andaman Sea, the deepest part is the central axial rift trending northeast–southwest and centered at 10.5°N: 94.5°E, with a depth of 4000 m (Curray et al., 1982). The rift valley can be traced to the south but is offset by NNW-SSE faults at places (Anon, 1981) (cf. Curray et al., 1982). The main sources of sediment to the basin are the Irrawaddy, Sittang, and Salween rivers in Myanmar. The Irrawaddy presently discharges about 265×10^6 tonnes of silty clay per year, 90% of which is deposited in the continental shelf (Rodolfo, 1969). During periods of high-level activity, the bulk of the sediment is trapped in the delta area or on the continental shelf. However, during low sea-level stands, they are introduced directly into the basin (Curray et al., 1982). The matrix data show that there is no direct input of sediment from the Bengal Fan into the Andaman Sea (Curray 1991; Roonwal et al., 1997a,b). The sediments in the Central Andaman Trough are principally turbidity flows (silty clays) with an overall sedimentation rate of about 150 mm/1000 years (Rodolfo, 1969). The Central Andaman Rift is both an active spreading center and a fan valley analogous to that in the prospective for hydrothermal minerals. The distal fan deposits have buried the rift valley in the north but have not penetrated to the southern part of the rift (Curray 1991, 1982; Anon, 1981).

6.7 Hydrocarbon Potential in the East African Coast

6.7.1 Sedimentary Basins in Somalia and their Hydrocarbon Resource Potential

On the east coast of Africa, Somalia is situated in the easternmost part of Africa and is known as the horn of Africa. Thus, Somalia occupies an important place along major sea transportation routes in the Middle East and Asia. Somalia is bounded by the Gulf of Aden on the north, Indian Ocean on the east, Kenya and Ethiopia on the west, and Djibouti on the northwest. Somalia's maritime area covers approximately 835,332 km^2 in size. The coastline runs over 3000 km, which makes Somalia the country with longest coastline in Africa. The northern mountains give way to plateaus (Oogo) and the plateau gradually drop into plains, which then steadily descend to the south toward the Indian Ocean. Perennial rivers, namely Juba and Shabelle, originating from Ethiopian high land flow through Somalia into the Indian Ocean and carry huge sediment (Figure 6.5).

In Somalia, the presence of sedimentary rocks is the primary element necessary for assessing hydrocarbon potential in the area. Throughout the geologic times, the Somali land has undergone several geological cycles during which tectonic activities created various patterns of uplifts, depressions, and rifting to produce the formation of sedimentary basins. The sedimentary basins of Somalia have very thick, over 8000 m of sedimentary sequence of various rock types that were deposited over different geological times.

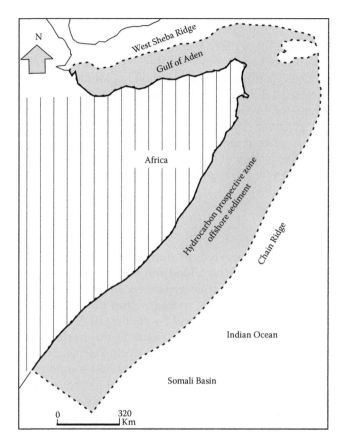

FIGURE 6.5
Offshore hydrocarbon potential zone along east African coast.

Based on the available information, all of the abovementioned sedimentary basins in Somalia have shown favorable geologic features comprising source rock, reservoir rock, and cap rock, and structural and stratigraphic traps for hydrocarbon accumulation. With the present-day cost-efficient technological advances in the oil and gas exploration industry, the African coast with potential hydrocarbon basins, which was in the state of frontier exploration phase for technical reasons, was explored and became a petroleum-producing region. Since there are technical innovations enabling the exploration of these basins, the major international oil companies are showing a renewed interest in Somalia's promising potential hydrocarbon basins.

6.7.2 Hydrocarbon Potential in Western Indian Ocean

The thick sedimentary succession along the east African coast makes the western Indian Ocean as an upcoming area of interest for hydrocarbon resources. This is supported by encouraging discoveries of oil and gas resources in the subregion, along the Mozambique to the Somalia coast. Gas discovery in Mozambique and Tanzania, oil deposits in South Sudan, offering of deep ocean exploration rights in Madagascar, Tanzania, and recently opened EEZ of Seychelles in the western Indian Ocean offer potential areas of hydrocarbons.

6.7.2.1 Mozambique

As mentioned above, the sedimentary succession along the African coast from western Indian Ocean seaboard had been there since the 1960s and Mozambican discovery was made in 2011. The renewed efforts for hydrocarbon explorations started in 2006 with the Mozambican state-owned oil company. This started the exploration activity of western Indian Ocean's nations for hydrocarbons. In 2007, the Norwegian state-owned energy firm joined the exploration race by setting its operational base in the Mozambican northern neighbor, Tanzania, when it signed a production sharing agreement (PSA) with the Tanzanian Petroleum Development Cooperation (TPDC). The US oil giant Anadarko joined the fray in the Mozambican exploration program, since 2010. Then the discoveries began.

6.7.2.2 Tanzania/Kenya

Natural gas deposits were discovered in Tanzania and Mozambique in the Indian Ocean. These discoveries boosted the added interest of Tanzania and Mozambique for more exploration while at the same time generated an intense interest across the western Indian Ocean region. The contagion spread within the Indian Ocean belt nations of Mauritius, Madagascar, Kenya, Somalia, and Seychelles. In April 2012, the US Geological Survey (USGS), prompted by the discoveries made in the western and southern Indian Ocean shelf, decided to explore further. Under its World Petroleum Resources Project, the USGS conducted a study within the region. The study had sought to assess the potential of undiscovered, technically recoverable oil and natural gas resources of this part of the world, exclusive of the United States. Seychelles, Comoros, Kenya, Tanzania, Madagascar, Mozambique, and South Africa were all covered in the assessment, which narrowed to four geological provinces within the western and southern Indian Ocean region.

6.7.2.3 Persian Gulf

The Persian Gulf represents one of the richest examples of hydrocarbon deposits both onland and offshore. Thick amount of sediments of Jurassic times containing organic material have been washed into the Gulf by the river system of the Tigris and Euphrates. The majority of the Indian Ocean, however, has a thin sediment cover in the range of 100–500 m and is generally not older than Eocene times, which eliminates hope in a major part of the ocean as a possible reservoir for methane deposits. However, along the coastal areas are some with great sedimentary thicknesses that can be considered for the exploration of petroleum, the most important being the northern part of the Arabian Sea. Several areas of large sediment thickness are found between Mozambique and Madagascar near Somalia. The richest of all is the neighbouring area of the Persian Gulf, the Sundra Shelf, and possibly the Red Sea are also important regions.

References

Anon, 1981. Andaman Sea—Gulf of Thailand (Transect IT). *Studies in East Asian Tectonics and Resources (SEATAEATAEATAR)*, CCOP, Bangkok, pp. 37–51.

Biswas, S. K., 1989. Hydrocarbon exploration in western offshore Basins of India. *Geol. Surv. India. Spl. Pub.*, 24: 184–194.

Biswas, S. K., 2008. Geodynamics of Indian plate and evolution of Mesozoic-Cenozoic basins. *Mem. Geol. Soc. India*, 74: 247–260.

Biswas, S. K., Bhasin, A. L., and Ram, J., 1993. Classification of Indian sedimentary Basins in the framework of plate tectonics. *Proceedings Second Seminar on Petroliferous Basins of India,* Indian Petroleum Pub. Dehradun, 1: pp. 1–46.

Biswas, S. K. and Singh, N. K., 1988. Western continental margin of India and hydrocarbon prospects of deep sea Basin. *Proceedings of 7th Offshore SE-Asia Conference*, Singapore OTSC 88119, pp. 170–181.

Chandra, K., Singh, R. P., and Julka, A. C., 1998. Gas Hydrate Potential of Indian Offshore Area. *Proc. 2nd Conf. & Exposition, Petroleum Geophys. Soc. Petrol. Geophys*, Dehradun, pp. 359–368, 560.

Chand, S. and Subrhmanyam, C., 2001. Subsidence and isostasy along a sheared margin—Cauvery Basin, eastern continental margin. *Geoph. Res. Lett.*, 28: 2273–2276.

Curray, J. R., 1991. Possible green schist metamorphism at the base of 22 km sedimentary section, Bay of Bengal. *Geol. South Pacific Geol. Notes*, 19: 1097–1100.

Curray, J. R., Moore, D. G., Lawver, L. A., Emmel, F. J., Raitt, R. W., Henry, M., and Kieckhefer, R., 1982. Structure, tectonics and geological history of the north-eastern Indian Ocean. In A. E. M. Nairn and F. G. Stehli (eds.). *The Ocean Basin and Margins: The Indian Ocean*. Plenum Press, New York, Vol. 6, pp. 399–450.

DGH, 2010. Petroleum exploration and production activities, India, 2009–10, Vol. 73, pp. 107–113.

DGH, 2015. Hydrocarbon exploration and production activities, India, 2014–15. pp. 14–17, 100–108, 112–118, 152–153.

Dwivedi, A. K., 2016. Petroleum exploration in India—a perspective and endeavour. *Proc. Indian Nat. Sci. Acad.*, 82(3): 881–903.

Rodolfo, K. S., 1969. Bathymetry and marine geology of the Andaman basin: Tectonic implications for Southeast Asia'. *Geol. Soc. Am. Bull.*, 80: 1203–1230.

Roonwal, G. S., Glasby G. P., and Chugh R., 1997a. Mineralogy and geochemistry of the surface sediments from the Bengal Fan, Indian ocean. *J. Asian Earth Sci.*, 15: 33–41.

Roonwal, G. S., Glasby, G. P., and Roelandts, I., 1997b. Mineralogy and geochemistry of surface sediments from the Andaman Sea and Upper Nicobar Fan, Northeast Indian ocean. *J. Indian Assoc. Sedimentol.*, 16(1): 89–101.

Roonwal, G.S., 2015. Competition for seabed resources in the Indian Ocean. National Maritime Foundation, New Delhi, Monograph, 101 pp.

Roy, S. K. and Das Sharma, S., 1993. Evolution of the Andaman Fore-arc basin and its hydrocarbon potential. *Proc. Second Sem., Petroliferous Basins of India*, 1: 407–435.

7

Living Resources: Fish and Fishery

7.1 Biomass for Fishery

Zooplankton comprise the most important ecological group of microorganisms, which is the food for fish growth in the sea. Because of this role between plankton and fish, the zooplankton are considered to be the chief index of utilization of aquatic biotope at the secondary trophic level (Goswami, 1996). Plankton studies from the Indian Ocean received impetus during the International Indian Ocean Expedition (IIOE, 1959–65). The data thus collected plankton atlases and handbooks (IOBC Atlases and Handbooks 1968–1973). Later studies of Qasim (1977), Peter and Nair (1978), Goswami et al. (1979), Goswami (1983), Nair et al. (1981), and Madhupratap et al. (1981) gave a quantitative distribution of zooplankton and zoogeography of common taxa of the oceanic realm. Although this was an important contribution, yet this sampling coverage was low and therefore only broad generalization could be arrived at. There is a need to conduct more studies for a better understanding of the potential of fisheries in the Indian Ocean.

Broad generalization on the assessment of zooplankton biomass information on secondary production and potential fisheries resources in the Indian Ocean was made between 1976 and 1991. The area investigated comprises the Arabian Sea, the Bay of Bengal, and the Laccadive Sea. The result showed that the shelf areas sustained higher zooplankton production averaged 0.62 ml m^3 compared to oceanic water average of 0.41 ml m^3. Biomass values were higher in the shelf region during the monsoon. The biomass secondary production and potential fish yield worked out to 9.01 million tonnes per year to 2.25 million tonnes per year respectively in this region (Goswami, 1996).

7.2 Tuna Fishing in the Indian Ocean

Tunas and tuna-like fishes are among the important fishery resources for most coastal nations as well as for distant-water fishing nations. The Indian Ocean contributes more than 1 million tonnes of the annual catch, which represents nearly 20% of the world production of tuna and tuna-like fishes. Because of the source of food and economic interest and the large distribution of major species, tuna is the focal point while addressing the issues of development, utilization, and management of fisheries in both EEZ regulations and international conventions. (Somvanshi et al., 1988; Sudarshan et al., 1989).

In the Indian Ocean, traditional tuna fishery has been done by several coastal nations. However, because of economic interest and good food value fishing in the Indian Ocean was carried out by Japan, Taiwan, and Korea. After United Nations Convention on the Law of the Sea (UNCLOS) and the EEZ regimes, the coastal nations began to increase their utilization of fish resources in their national jurisdiction. They made indigenous fisheries net and also partnered through licensing arrangements with other fishing nations. The emergence of purse-seine fishery in the western Indian Ocean during 1982–83 by French and Spanish fleets was a major milestone in the development of Indian Ocean tuna fishery. The increasing effort by the coastal states and distant-water fishing nations led to a rapid growth of Indian Ocean tuna fishery. It led to high growth of 286% in production during the last two decades compared to a growth of 47% and 113%, respectively, in the Atlantic and Pacific Ocean fishing. Silas and Pillai (1982), Stequert and Marsac (1989), and Sudhakara (2009) have reviewed the tuna fisheries of the Indian Ocean. The issues concerning tuna fisheries in the countries participating in the fishery were reviewed by FAO/IPTP Experts. Consultations on the Indian Ocean tunas were held at Seychelles during October 1993 (Ardill, 1994) and Colombo during September 1995 (Anon, 1995a) and today there is an Indian Ocean Tuna Commission. The current status and recent trends in the Indian Ocean tuna fishery, with emphasis on the oceanic tunas, are as follows.

7.3 Fish Catch Status

The annual production of tunas and tuna-lime fishes in the Indian Ocean has reached the high level of 1.3 million tonnes, comprising oceanic tunas (79%), neritic tunas (17%), and billfishes (4%).

Among the oceanic tunas, contributing to about 0.69 million tonnes, the yellowfin tuna (*Thunnus albacores*) and shipjack (*Katsuwonus pelamis*) are the major components yielding 0.35 million tonnes an 0.25 million tonnes respectively, which represent 50.7% and 36.3% of the oceanic tuna production. Bigeye tuna (*T. obesus*), albacore (*T. alalunga*), and southern bluefin (*T. maccoyii*) are the other species forming 9.8%, 2.6%, and 0.6% of the catches respectively. The trend in the production of oceanic tunas is given in Figure 7.1.

The neritic tunas yielding about 1.15 million tonnes annually are represented by five species, viz., kawakawa (*Euthynnus affinis*), longtail tuna (*Thunnus tonggol*), frigate tuna (*Auxis thazard*), bullet tuna (*A. rochei*), and Indo-Pacific bonito (*Sarda orientalis*) in their order of abundance. Within these, the prominent species are kawakawa and long-tail tuna, accounting for 36.4% and 22.3%, respectively, of the neritic tuna catch.

The billfishes, contributing to about 35,000 tonnes of catch, annually consist of the blue marlin (*Makaira mazara*), black marlin (*M. indica*), striped marlin (*Tetrapturus audax*), Indo-Pacific sailfish (*Istiophoru platypterus*), and swardfish (*Xiphias gladius*). These species come mostly from longline fishery.

7.3.1 Tuna Production and Fishery Type

The commercial fishing methods for the tunas have been classified as (a) surface and (b) subsurface fishing methods. While the former largely includes purse-seining, pole and line fishing and drift gillnetting, subsurface fishery is done by long-lining. In the oceanic tuna fishery in the Indian Ocean, purse-seining and long-lining are the principal fishing methods, accounting for 40.4% and 32.3% of the catches respectively. The species-wise catch taken by each fishery is given in Table 7.1.

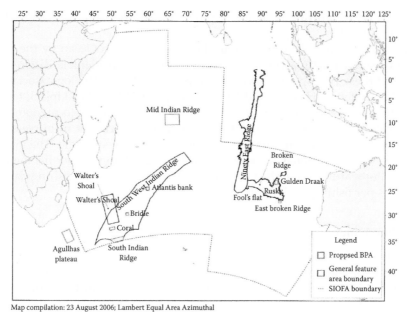

Map compilation: 23 August 2006; Lambert Equal Area Azimuthal

FIGURE 7.1

Fishing zones in the southern Indian Ocean. (From www.siodfa.org)

7.3.2 Purse-seine Fishery

Large-scale industrial purse-seining commenced in 1982–83. French and Spanish fleets from the tropical Atlantic entered the eastern Indian Ocean for fishing. Presently, a large number of vessels from Spain, France, Russia, Brazil, Panama, Mauritius Japan, Thailand, and Taiwan fish in the Indian Ocean. Most of the vessels are of >1000 GRT class. The fishery is developed around Seychelles and it is most active between latitudes 10°S and 5°N.

The annual catch is large, covering the tropical Indian Ocean due to high productivity in this region. The amount of catches taken from different geographical regions is indicated in Figure 7.1. During the developmental phase of the fishery, the production went up steadily and reached the peak level of more than 3 million tonnes in 2012. The catches have more or less stabilized since then, though there has been a marginal drop in the fishing

TABLE 7.1

Fish Type and Catch

Species-Wise Catch (in '000 tonnes)						
Fishery Type	Yellowfin Tuna	Bigeye Tuna	Albacore	Southern Bluefin	Skipjack	Total
Longline	148.6	54.9	16.5	2.6	0.5	223.1
Purse-seine	136.1	13.0	1.1	0.5	128.1	278.8
Gill net	28.0	–	–	–	38.0	66.0
Pole and line	10.0	–	–	1.2	64.1	75.3
Troll line	4.1	–	0.1	–	4.7	8.9
Handline	1.4	–	–	–	2.5	3.9
Unclassified	21.7	7.0	–	–	12.6	34.3
Total	349.9	67.9	17.7	4.3	250.5	690.3

effort, because of the shifting of Japanese purse-seiners from the western to the eastern Indian Ocean since then.

The major components of the catches in the western Indian Ocean by the purse-seine fishery are of yellowfin tuna (49%) and shipjack (46%). Considerable differences in the proportions of these two species have been reported between the log-associated catches and free-school catches (Hasting and Domingue, 1995). While the skipjack tends to predominate in the log-associated catches (59%–75%), yellowfin tends to predominate in the free-school catches (50%–64%). Bigeye tuna is occasionally caught in purse-seining and represents 3%–5% of the catch. The catch of albacore is very sporadic. The average catch per unit effort (CPUE) of tunas recorded in 1994 was 22.32 tonnes per fishing day. There was a marked upward trend in the CPUE from about 13 tonnes per fishing day in 1984 to over 22 tonnes per fishing day in 1994.

7.3.3 Longline Fishery

Longline fishery, which started in the eastern sector in 1952, moved gradually across the tropical belt of the Indian Ocean and then expanded toward the north into the Bay of Bengal and the Arabian Sea. Finally, the fleets moved southward and these now cover virtually the whole of the Indian Ocean. Taiwan, Japan, and Korea are the major fishing nations operating for this fishery with fleets of 308, 180, and 50 vessels, respectively. The Indian Ocean countries like Indonesia, Iran, Pakistan, and India have also entered the fishery with the introduction of ownership vessels and by licensing arrangements with distant-water fishing nations. In recent years, large catches, mostly of yellowfin tunas, are being harvested from the northern Arabian Sea.

The annual catch of tunas, which was in the range of 131,000–155,000 tonnes during 1990–1992, went up to a remarkably high level of 223,000 tonnes recently, largely due to the increase by the Taiwanese vessels. The catch consisted of yellowfin tuna (66.6%), bigeye tuna (24.6%), albacore (7.4%), and southern bluefin (1.2%).

7.3.4 Pole and Line Fishery

Pole and line fishery is practiced mainly in the Maldives, the Lakshadweep Islands (India), and Australia. This small-scale fishery contributes to about 75,000 tonnes of annual catch of which over 68,000 tonnes is landed in the Maldives. While skipjack (85%) and yellowfin (15%) are the catch components in the Maldives and Lakshadweep, southern bluefin is the largest resource in the Australian pole and line fishery.

7.3.5 Gillnet Fishery

High-seas gill netters, which were operated in the Indian Ocean from 1986 to 1991, targeted mainly southern bluefin and albacore. This fishery ceased to exist in response to UN General Assembly Resolution 46/215 in the year 1991. Presently gillnetting is only a small fishery and accounts for a total catch of 66,000 tonnes of skipjack and yellowfin tuna.

7.3.6 Troll Line and Handline Fisheries

Troll line and handline fishing, accounting for 10,000 to 15,000 tonnes catch annually, are prevalent in Comoros, Lakshadweep, and Maldives. Skipjack and young yellowfins are the

TABLE 7.2

Depth-wise Fishery Resources in the Indian Ocean

Resource	Depth			
	0–50 m	50–500 m	Oceanic	Total
Demersal	1.280	0.653	–	1.933
Pelagic	1.000	0.742	–	1.752
Oceanic	–	–	0.246	0.246
Total	2.280	1.395	0.246	3.921

main species caught. Since the introduction of EEZ of 200 nautical miles, the marine fishing world over including the Indian Ocean has seen many changes in strategy for fishing.

The aim being optimum fishing and concentration as proposed by several countries at present, the coastal zones up to a depth of 50 m are under stress of overfishing and exploitation by both traditional and mechanized boats. Consequently, there is no possibility of further catch from this depth. However, fish resource potential beyond 50 m is possible; this is summarized in Table 7.2.

Demersal resources comprise total stock of 1.93 million tones, coastal zone upto a water depth of 50 m, and support 1.28 million tonnes, whereas the water depth beyond 50 m is predicted to yield 0.633 million tonnes as can be seen from Table 7.2. The variety of fish catch in such case consist primarily of threadfin breams, mackerel, bull eye, scads, ribbon fish, and Sciaenids. Considering the catch level, it is possible to conclude that there exist a possibility to increase more fishing in the region beyond 50 m water depth.

7.4 Tuna Fish in the Indian Ocean

Tuna, hill fish, myctophid, and oceanic squid spices are available for further fishing, of course, within the recommendation of Indian Ocean Tuna commission (Sudarshon et al., 1989; IOTC 2010, 2011), which has estimated the potential of Ocean Pelagic fish resources from subsurface fishing.

Conservation and management of fish resources would depend upon the quantity of fish resource available for fishing and the level of quantity of fishing for a sustainable level. A code of conduct for responsible fishing as proposed by FAO in 1995 is relevant. This focus of the coastal states on their own responsibility for the development and adoption of technology to prevent catch of nontarget species thus prevents wastages through discard. This will greatly reduce threat to endangered species and increase survival by escaping fishing.

7.5 Some Challenges of Fishing in the Indian Ocean

The Indian Ocean and its adjacent seas have the largest number of active fishermen in the world. And nowhere else have so many fishermen been killed, fired on, or arrested. According to press reports, over the last decade, about 200 fishermen have been shot dead in the region, and hundreds injured. In the last four years, about 1500 fishermen have been arrested. Some of them have languished in foreign jails for up to three years. This takes place despite the fact that imprisonment, as a penalty for violating fishery regulations, is contrary to the provisions of the 1982 UNCLOS.

While death by firing is mainly confined to the waters between India and Sri Lanka, arrests and detention are reported from all over the Indian Ocean region—Indians by Somalia, Pakistan, Sri Lanka, Maldives and Bangladesh; Pakistanis by Iran and India; Myanmarese by Bangladesh and India; Eritreans by Yemen from the Red Sea to the Arabian Sea, from the Bay of Bengal to the Indian Ocean, there are hundreds of cases of fishermen being arrested and detained by authorities for undertaking fishing activities or for just being found drifting in alien waters (Sampat, 1998; Somvanshi, 2001).

It is disturbing to note that these incidents are that these are just small-scale fishermen working on board vessels that are 10 to 12 meters in length, with barely enough space to stretch their back, some working for a wage, and some for a share of the catch. It needs to be understood that this is a problem of human survival for livelihood. Otherwise why would a fisherman illegally cross the international maritime boundary, risk arrest and detention, and sometimes even face arrest and shoot out?

1. Fishermen get arrested out of sheer ignorance about maritime boundaries. There are no easy ways to demarcate boundaries at sea, unlike on land. Unless one knows coordinates—for which you need navigation equipment—one might end up in neighboring waters and get arrested. It also needs to be understood that fishermen accidentally drift into neighbouring waters because of engine failure or strong currents.

2. Some maritime boundaries are unresolved, as between India and Pakistan in the Sir Creek area. Fishermen have no clue if they have the right to fish where they would like to, especially in fishing areas in contested waters. Even if fishermen from both sides do not have any problem in sharing a fishing ground. this seems to be the case between Karachi and Porbander fishermen, for instance, the coast guards often arrest them.

3. Fishermen cross over because they think it is morally right to continue to fish in areas where they had enjoyed traditional fishing rights. Indians, for instance, have traditionally fished around Katchatheevu in the Palk Strait, which is today part of Sri Lanka. Eritreans and Yemenis have traditionally fished in the waters of Zuqar-Hanish and Zubayr groups of islands in the Red Sea, which are now divided between Eritrea and Yemen.

4. The important fishing in other exclusive economic zones (EEZs)—the marine space within which the coastal state has exclusive right to living and nonliving resources—occurs because of too many boats chasing too few fish in national waters. Take, for instance, the Rameswaram trawlermen crossing over to Sri Lankan waters, braving bullets. They take that risk since the shrimp fishing grounds on the Indian side are depleted, while rich shrimp resources exist on the Sri Lankan side. Similarly, Sri Lankans from the south of the country, mainly from Negombo, go all over the Indian Ocean, prompted by the lack of productive fishing opportunities in their waters and the abundance of tuna resources in other EEZs.

Regional coordination is vital to combating illegal poaching, unreported and unregulated (IUU) fishing. Fishing vessels are expanding their fishing areas well outside waters of their own countries onto the high seas and into distant EEZs. Identifying and catching IUU vessels requires advanced technology and a strong labor force. This is a daunting task for individual countries, particularly developing ones. However, with regional cooperation and information sharing, fighting IUU fishing becomes much more achievable. Secure Fisheries therefore works to reduce IUU fishing both within Somali waters and across the greater Indian Ocean region given below are some important parameters:

1. Halting IUU fishing is imperative for the international community. IUU fishing is an enormous problem that costs the global economy up to $23 billion a year. In the Indian Ocean, a significant amount of catch qualifies as IUU: 18% in the western Indian Ocean and 32% in the eastern Indian Ocean. Understanding the magnitude of IUU fishing and the necessity for an international solution, the Food and Agriculture Organization of the United Nations (FAO) created the port state measures agreement (PSMA). The PSMA is an international agreement that combats IUU fishing through more stringent port controls and exchange of information. It aims to eliminate ports of noncompliance where IUU vessels can land their catch undetected. The treaty was entered into force on June 5, 2016.

2. Secure Fisheries has joined with the United Nations Office on Drugs and Crime (UNODC) to facilitate the meeting of the Fisheries Crime Group of the Indian Ocean Forum on Maritime Crime (IOFMC). The forum provides a regional network for Indian Ocean states to coordinate their responses to maritime crime concerns at the strategic and operational levels. The IOFMC Fisheries Crime Group is a voluntary initiative with the following goals:

   a. Strengthening port controls to reduce safe havens for illegal fishing vessels.

   b. Increasing the use of automatic identification systems (AIS) on fishing vessels to achieve parity with merchant ships. Strengthening legislation of member states in order to criminalize illegal fisheries crime. Working toward mandatory, unique, and permanent ship identification numbers that meet the standards of the International Maritime Organization (IMO) to better identify illegal fishing vessels.

   c. Secure Fisheries advises on the above issues and actively promotes efforts to improve coordination and access to information between national, regional, and international programs. While the forum is mainly focused on illegal fishing, achieving the IOFMC Fisheries Working Group goals will aid in the reduction of unreported and unregulated fishing activities through data sharing and increased seafood traceability, thus aiding the overall advancement of sustainable fisheries.

7.6 Fishing in the Southern Indian Ocean

Commercial fishing in the southern Indian Ocean began in 1994 for a better and SDG utilization of fish resources. Here, a southern ocean fishing agreement was concluded.

The Southern Indian Ocean Fisheries Agreement (SIOFA) is aimed at ensuring the long-term conservation and sustainable use of fishery resources (excluding tuna) that fall outside national jurisdictions. SIOFA has 11 signatories and parties to the agreement and is soon due to be ratified. This will be a major step forward as the Regional Fisheries Management Organization (RFMO) covers areas of the high seas where no such organization or arrangement previously existed. For more information on the SIOFA fishing zone, the Southern Indian Ocean Deepsea Fishers Association (SIODFA) website can be accessed at www.siodfa.org (Figure 7.1).

7.7 Fishery in the Mangroves of Indian Ocean

Mangrove ecosystem is special environment since they include territorial plants growing in marshy or swamps contain saline water. They have a dense roots system and a network of pathways. The tree or vegetation produces enough organic matter, which flows into the water around to give rich food to the marine organism.

The significance of mangroves in the local fishery is seen that a variety of fin fish exists in the swamps. Several species from the mangroves of the Indian Ocean are known (Jayaseelan, 1998). In a country like Malaysia, more than 30% of fish are captured is in the mangrove region and on the west coast, it is as high as 40%. In the Bay of Bengal in Sunderbans of India and Bangladesh, its influence is pronounced. The entire Indian Ocean region is favorable for shrimp catch from the mangrove–delta zone of India, Thailand, Malaysia, Indonesia, and several along East African coastline.

7.8 Fishing in the Indian EEZ

Within her EEZ, the vast majority of effort takes place within inshore waters. Of this mainly 6.2% of the catch is by traditional nonmechanized methods. However, there is now an awareness of the requirement for more mechanized boats, although finance is always the main consideration. In a recent survey, it was estimated that the number of mechanized boats now total of 19,000 of which 11,600 are trawlers and 3,990 gill netters, 2,850 bagnetters, and 380 purse seiners. This compares with 140,000 nonmechanized boats (Infofish Mkt Digest).

In spite of the fact that government support is assured, the importance of the fisheries industry varies from state to state along eastern and western coasts. In spite of all the efforts of the government to make fish an export item, most of the catch is consumed fresh, either by the fishing population or local markets. This effect is exacerbated by a poor infrastructural facility for storage and packing. Nevertheless, the government is keen to develop offshore fishing resources, and the procurement of a Danish-aided modern fishing vessel *Sagar Sampada* is a pointer to this. Since 1950 fish production has increased, but growth still does not match the desired production level. Also from the international market perspective, it is at a low level.

Although India has access to vast quantities of living resources, in terms of fisheries, and aquaculture/mariculture, they have remained relatively underexploited. There have been attempts, however, both through the Indian Council of Agriculture Research, and Ministry of Earth Sciences, to extend full government support to increase the yield and modernize the fishing industry. Both fisheries and aquaculture/mariculture are important

TABLE 7.3

Profile of Indian Marine Fisheries

Physical Component	Profile	
	Length of coastline	8129 km
	Exclusive Economic Zone	2.02 million km²
	Continental shelf	0.50 million km²
	Inshore area (<50 m depth)	0.18 million km²
	Fishing villages	3288
Human Component		
	Marine fisher population	4.0 million
	Active fisher population	0.99 million
	Fishermen families	0.86 million
Infrastructure Component		
	Landing centers	1511
	Major fishing harbors	6
	Minor fishing harbors	27
	Mechanized vessels	72559
	Motorized vessels	71313
	Nonmotorized vessels	50618

and significant for future development, especially in regard to the exclusive economic zone (EEZ).

Resource assessment carried out on pelagic fishes indicate potential catches as follows: along the western coast alone, oil sardine (400,000 tonnes), mackerel (300,000 tonnes), horse mackerel (150,000 tonnes), and catfish (150,000 tonnes). Of the pelagic fishes, oil sardine and mackerel are fairly well exploited. Other species exploited include tuna fish, the frigate mackerel, and bonitos. There also exists appreciable potential for increased squid fishing (Table 7.3) (James, 2014; Sathianandan, 2015).

7.9 Fisheries Management and Best Practices in the Indian Ocean

The coastal states in the Indian Ocean share a keen interest in the management and conservation of the region's rich fish resources. Fisheries and related industries are critical in ensuring food security. The impacts of overfishing and climate change have accelerated the reduction of major fish stocks in the region. The newly established Fisheries Support Unit (FSU), hosted by the Sultanate of Oman, manages and spearheads IORA efforts to identify and discuss key fisheries-related issues mentioned in the action plan. It also serves to study proposals and facilitate research in areas that are of practical use to member states. Figure 7.2 show fish type in the Indian Ocean.

The FSU acts as a regional center for knowledge sharing, capacity building, and addressing strategic issues in the fisheries and aquaculture sectors. In order to tackle the challenges facing regional fisheries management, member states should expand on this cooperative mechanism. In addition, it might be appropriate to find ways to remove barriers and

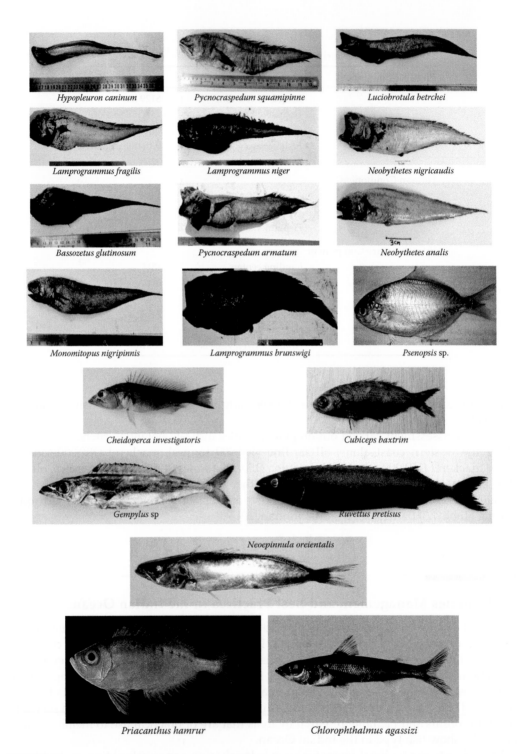

FIGURE 7.2
Assemblage of different types of fishes in the Indian Ocean. Maxi size attained by many of these fishes 1–1.2 m. (Courtesy: Centre for Marine Living Resources and Ecology (MOES.)

boundaries between separate parts of the fisheries management process including science, policy and decision-making, and stakeholder engagement, so as to bring them together into regional management forums.

The role of the FSU as an Advisory Committee for Fisheries and Aquaculture may be further explored. It has been reported that the future global demand for fish consumption will have to be met by aquaculture and, therefore, there is a need for further research and development in this sector. IORA may also consider developing regional/international networking to interact with other institutions sharing common interests in fisheries management.

Fisheries management should be at the core of the new maritime policy that the IORA should consider developing as there is a need to build mutual understanding among all decision-makers and players of the maritime industry. Effective decision-making must also integrate environmental concerns into maritime policies as maritime pollution plays a major role in the decline of fish stock.

Marine fisheries in India is a multispecies fishery. Around 1400 finfish species are harvested from the sea of which 263 are commercially important. Apart from this, 36 species of penaeid shrimps and 34 species of cephalopods are also harvested in which 15 species of penaeids and 8 species of cephalopos are commercially important (ICAR Report, 2013; CMFRI Annual Reports).

The marine fishery resources from the Indian seas are harvested using more than 35 different types of craft gear combinations. The major crafts used are of three different categories namely mechanized, motorized, and nonmotorized. The mechanized sector includes trawlers, gill-netters, and inboard vessels. Most of the crafts in the mechanized sector use machines for both propulsion and operation of the gear. The motorized sector exclusively consists of crafts fitted with outboard engines. The nonmotorized sector consists of traditional vessels made up of wood, fiber glass, thermocol, and so on, and do not use any machine power either for propulsion or for operation of the gear. Major gears used in the marine fisheries sector are trawl nets, gill nets, bag nets, hooks and lines, and seines (Vivekanandan, 2006; CMFRI Annual Reports).

According to the present status, the largest buyer of Indian marine products is South East Asia with 39.9% share in volume and 25.09% share in value (US$). The next highest buyer is European Union with 22.96% share in volume followed by the United States (18.17%), Japan (13.01%), China (7.51%), Middle East (5.33%), and 7.5% (to other countries). Export to South East Asia recorded a growth of 45.01% in volume and 87.51% in US$ realization. This is mainly due to the increased export of frozen shrimp, frozen fish, and chilled items. Exports to United States registered a growth of 36.45% in quantity and 45.39% in value (US$ realization) and this is mainly due to increased export of frozen shrimp and cephalopods.

Exports of Vannamei shrimp showed a tremendous increase in the US market by 212% in quantity and 209% in US$ realization. Export to Japan also registered a positive growth of 21.33% in quantity and 22.35% in US$ terms. Exports of chilled items showed a tremendous increase in Japanese market by 120.12% in quantity and 220.34% in US$ realization. Exports to China showed a drastic decline of 46.89% in quantity and 40.17% in US$ terms. The marine products exports have strengthened India's presence in South East Asia. There is a significant increase in exports to South East Asian countries compared to the previous year. Export of frozen shrimp to South East Asia has registered a growth of about 222.43% in volume and 356.36% in US$ terms. Export of frozen shrimp to the United States has also showed a growth of about 47.68% in volume and 47.55% in US$ terms. Export of Vannamei shrimp had also picked up. We have exported about 40787 MT of Vannamei shrimp during this period. Export to Middle East countries showed an increase of 25.98% in US$ realization but declined in quantity by 13.25%.

References

Anon, 1995a. Report of the Expert Consultation on Indian Ocean, *6th session*, Columbo, Sri Lanka, 25-29 September 1995: IPTP: 67 pp.

Ardill, J. D. (ed.), 1994. *Proceedings of the Expert Consultation on Indian Ocean Tunas*, 5th Session, Mahe, Seychelles, 4–8 October 1993. 275 pp.

FAO, 2005. Aquaculture production 'Year Book of Fisheries Statistics, Vol.96.

Goswami, S. C., 1983. Zooplankton incidence in abnormally high tea surface temperature in the Eastern Arabian Sea. *Indian J. Marine Sci.*, 12: 118–119.

Goswami, S. C., 1996. Zooplankton biomass and potential fishery resources in the EEZ of India. In: S. Z. Qasim and G. S. Roonwal (eds.). *India's Exclusive Economic Zone*, Omega, New Delhi, pp. 94–104, 238.

Goswami, S. C., Selvakumar, R. A., and Goswami, U., 1979. Diel and tidal variations in zooplanktonic populations in the Zuariestuary, Goa. *Mahasagar-Bull. Natn. Inst. Oceanography*, 12(4): 247–258.

Hasting, R. E. and Domingue, G., 1995. Recent trends in Seychelles industrial fishery. *FAO/IPTP Expert Consultation on Indian Ocean tunas*, Colombo, Sri Lanka, 25–29 Sept. 1995.

Indian Council of Agricultural Research (ICAR), 2013. Handbook of fisheries and aquaculture reports.

International Indian Ocean Expedition (IIOE), 1959-65. Various reports.

IOTC, 2010. Report of the twelfth session of the IOTC Working Party on Tropical Tunas (WPTT), Victoria, Seychelles 18–25 October 2010 -WPTT-R[E], 82 pp.

IOTC, 2011. Report of the thirteenth session of the IOTC Working Party on Tropical Tunas, Republic of Maldives, 16–23October 2 WPTT13–R[E], 94 pp

James, P. S. B. R., 2014. Deep sea Ashing in the Exclusive Economic Zone of India - resources, performance and new - approaches to development. In: S. A. H. Abidi and V. C. Srivastava (eds.). *Marine Biology*. The National Academy of Sciences, India, Allahabad, pp. 100–123.

Jayaseelan, M. J. P., 1998. Manual of fish eggs and larvae from Asian mangroves waters. *UNESCO*, 193 pp.

Madhupratap, M., Achuthankutty, C. T., Nair, S. R. N., and Nair, V. R., 1981. Zooplankton abundance of the Andaman Sea. *Indian J. Marine Sci.*, 10: 258–261.

Nair, S. R. S., Nair, V. R., Achuthankutty, C. T., and Madhupratap, M., 1981. Zooplankton composition and diversity in western Bay of Bengal. *J. Plankton Res.*, 3: 493–508.

Peter, G. and Nair, V. R., 1978. Vertical distribution of zooplankton in relation to thermocline. *Mahasagar—Bulletin of the National Institute of Oceanography*, 11:169–75.

Qasim, S. Z., 1977. Biological productivity of the Indian Ocean. *Indian J. Mar Sci.*, 6: 122–137.

Sampat, V., 1988. Sustainable fisheries and environment. In: S. Z. Qasim and G. S. Roonwal (eds.). *Living Resources of the Indian EEZ*. Omega, New Delhi, pp. 65–85, 140.

Sathianandan, T. V., 2015. Status of marine fishing resources in Indian Ocean—An Overview, CMFRI, 11–22.

Silas, E. G., and Pillai, P. P., 1982. Resources of tunas and related species and their fisheries in the Indian Ocean. *Bull. Cent. Mar. Fish. Res. Inst.*, 32: 174.

Somvanshi, V. S., 2001. Problems and prospects of deep-sea and far-sea fishing. In: T. J. Pandian (ed.). *Sustainable Indian Fisheries*, NAAS, New Delhi, pp. 71–87.

Somvanshi, V. S., Kedian, A. S., and John, M. E., 1988. Present status of Tuna fisheries in the Indian Ocean. In: S. Z. Qasim and G. S. Roonwal (eds.). *Living Resources of the Indian EEZ*, Omega, New Delhi, pp. 18–29, 140.

Stequert, B. and Marsac, F., 1989. Tropical tuna—surface fisheries in the Indian Ocean. FAO Fish. Tech. Pap., No 282, Rome, FAO, 238 pp.

Sudarshan, D., Sivaprakasam, T. E., Somvanshi, V. S., and John, M. E., 1989. Assessment of oceanic tuna and allied fish resources in the Indian EEZ. *Proceedings of the National Conference on Tunas, CMFRI*, Cochin, p. 44–66.

Sudhakara Rao, G., 2009. *Deep-Sea fisheries of India*. B.R. Publications, New Delhi, 524 pp.

Vivekanandan, E., 2006. Oceanic and deep-sea fisheries of India. In: *Handbook of Fisheries and Aquaculture*. Indian Council of Agricultural Research, New Del/zf, pp. 95–105.

8

Energy and Fresh Water from the Ocean

8.1 Ocean Energy and Types

Before the commencement of the IIOE, practically nothing was known about the causes and pattern of the reversing wind system over the Indian Ocean. There existed an urgent need to investigate even in general terms such features as the structure of the equatorial current and Somali current. Likewise, almost nothing was known about the salinity and temperature gradient in the water column. IIOE made significant contributions in many areas, particularly the generation of data concerning the horizontal and vertical distribution of temperature, dissolved oxygen, salinity, and nutrient content in the waters. This led to the identification of the different water masses in the Indian Ocean and the sources of influence upon them. More specifically the data collected on the Bay of Bengal and the Arabian Sea led to a basic understanding of coastal dynamics along both the eastern and western coasts Narin and Stehli (1982). The other important contribution could be counted as the identification of undercurrents in the equatorial region. All these data help to understand the structure and dynamics of the monsoon pattern, on which India is so dependent in every sense. There are at least six methods by which power can be produced from the ocean (Cohen, 2009; Khan et al., 2009).

During the past few decades, nations were concerned by global warming and air quality (pollution) is getting disturbed due to coal burning for producing energy. The oil crisis of the past decades rendered this problem even more a serious to plan for alternative and clear source of energy. In this respect, the use of nonpolluting renewable from of energy received attention. This is important to adopt as early as possible to protect air quality from further damage.

It is known that solar radiation not only sustains life but gives it inexhaustible source of energy. It is broadly calculated that 10^{16} W of solar energy reaches the earth's surface. This is a natural collection of energy in the ocean water. Therefore, oceans have enormous potential to provide energy in various ways. The major advantage of energy produced from ocean is it is environmental friendly and thus nonpolluting and therefore has minimum health hazards. It is hoped that in the coming decades ocean energy shall become an important form of alternative energy source. This is more so for remote islands situated in the Indian Ocean.

The ways and form in which ocean energy could be produced and tapped and comprise (a) ocean thermal energy conversion (OTEC), (b) wave energy, (c) tidal energy, (d) offshore wind energy, (e) ocean current energy, (f) ocean salinity gradient energy, and finally (f) marine bioconversion energy (Table 8.1) (Indian Wave Energy, National Institute of Ocean Technology).

TABLE 8.1

Estimates of Indian Ocean Energy Resources

Type	Capacity (MW)
Ocean thermal energy (OTEC)	50,000
Wave energy	60,000
Tidal energy	8,000

Source: Raju, V. S. and Ravindran, M., 1985. Ocean energy in the Indian context, Mahasagar 18, No 2. National Institute of Oceanography, pp. 211–17.

8.2 The Monsoon Pattern and Thermal Structure

The ISMEX Programme (1973) investigated the response of northwest Indian Ocean to the monsoon wind system and also aimed toward investigating the thermal structure of the waters. It was thus realized that a decrease in the sea surface temperature (SST) was associated with a deepening of the surface layer. Subsequently the MONEX Programme (1979), organized as a part of the first global experiment (GARP), resulted in additional data on the central Arabian Sea. Additional surveys were aimed at investigating the cooling phenomenon of the Arabian Sea with the advance of the southwest monsoon winds. The remarkable change in the thermal structure as well as circulation in the upper 500 m of the Arabian Sea with the advancement of the monsoon appears to be associated with the change in the vertical sea column with the westward jets at 200 and 800 m depth near the equator.

India has recently made attempts at ocean modeling. If India plans to explore the ocean resources in the Indian Ocean, a rational modeling study of the oceanographic parameters is required, and as far as India is concerned, just a beginning has been made.

8.3 Ocean Thermal Energy Conversion

The ocean thermal energy conversion (OTEC) uses the thermal gradient of the ocean to utilize in generating heat engine to produce energy, or work output. It is generally seen as a simple concept known for several decades. However, the increasing demand for energy, especially after the "oil crises" technologists and nations look for alternative source of energy. In the OTEC, there are technology question to be able to utilize the potential of OTEC system (Vega Luis, 1992). Now that this part is largely mastered, attempts are being made to produce OTEC energy on commercial basis (Figure 8.1).

It has been mentioned earlier that there are several technological challenges to demonstrate for larger ratings of OTEC plants as an economical and commercial venture. India, being an energy-deficient nation, aims to harness OTEC energy since India has large coastline and the tropical climatic pattern of the sea water helps in realizing OTEC energy. In this regard, attempts are being made by the National Institute of Ocean Technology (NIOT) toward successful OTEC utilization and power generation.

To achieve these objectives, the NIOT has taken up the challenge of fabricating a floating 1 MW OTEC plant mounted on a barge in a 1000 m water depth about 40 km off Tuticorin

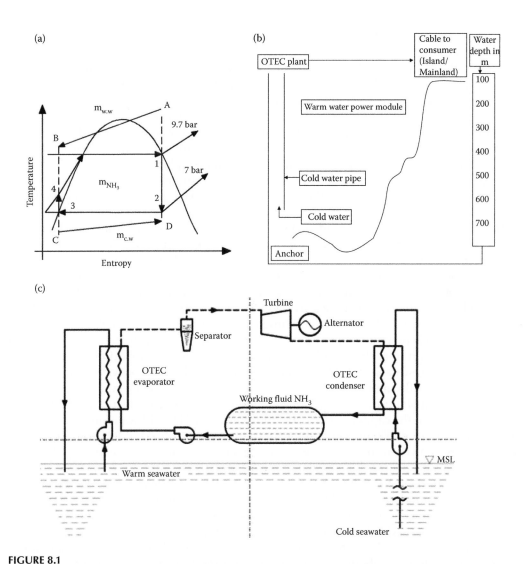

FIGURE 8.1
(a) Principle of simple Rankine cycle based OTEC (Courtesy: NIOT), (b) OTEC concept explained—the operating temperatures regime of the ocean. The power thus produces transmitted in a cable to the Island/mainland (consumer) territory, and (c) fabrication of simple Rankine cycle-based OTEC. (Courtesy: NIOT.)

in Tamilnadu. In doing so, the major challenge was the design of the platform and the cold water pipe. NIOT designed a non-self-propelled barge to suit the purpose with special features such as three moonpools and a retractable cold water sump to suit the NPSH requirements of the pumps. The barge was built in a shipyard on the west coast of India and was named "Sagar Shakthi."

Subsequently the same concept of using thermal gradient has been used for desalination successfully in various configurations in islands as well as offshore in over 500 m water depth. A 100 m³/day land-based plant was commissioned at Kavaratti Island in Lakshadweep in 2005. This plant has been continuously generating fresh water to meet the drinking water needs of the island community. LTTD plants, each with 100 m³/day capacity, have also established at Minicoy (2011) and Agatti (2012) islands and they have been operational since then.

NIOT has also successfully demonstrated usefulness of LTTD principle for waste heat recovery in coastal thermal power stations that discharge huge amount of condenser reject water into the nearby sea. This demonstration plant cum experimental facility is set up in North Chennai Thermal Power Station (NCTPS). NIOT is now focusing its efforts on powering desalination process using OTEC. Toward this aim, a laboratory is being set up to run OTEC and desalination cycle together. A small-capacity OTEC turbine with R134a as working fluid has been developed for the first time ever in-house at NIOT and more developmental studies on turbines and other important areas are in the offing. This laboratory will pave the way for prototype OTEC-driven desalination plants.

Geographically, India is well situated as far as ocean thermal energy potential is concerned. Around 3000 km of the southern Indian coast has a temperature difference between the surface and deep water of 20°C throughout the year. Thus, an area of 300,000 km² exists within the EEZ of India with an adequate temperature gradient. In addition to the Indian coastline, potential sites exist around the Andaman Nicobar and Laccadive Islands.

A feasibility study by the Ocean Engineering Centre at Madras also involved the public companies Bharat Heavy Electricals, Engineers India, and the National Institute of Oceanography in Goa. This study focused on the suitability of fabricating a 1 MW OTEC plant in the Laccadives, which are a string of 36 islands off the southwestern coast. Thus in 1982, the two islands, Kavaratti and Minicoy, were identified as possible sites. Detailed bathymetry and temperature surveys were conducted showing that a depth of 1000 m occurs about 31 km from the islands accompanied by a temperature difference of 20°C to 22°C between the surface and 1000 m depth throughout the year. Surveys indicated also that the subsea bathymetry of Kavaratti is more suitable for the installation of the cold water pipe, being of uniform slope. It was therefore proposed to build a 1 MW OTEC pilot plant at Kavaratti, the capital of these groups of islands. The cold water intake was to be located at a depth of 650 m where the water temperature averages 10°C (DOD Annual Report, 1986). However, this project was subsequently rejected as the over population of the islands (50,000 people, 32 km² island) meant insufficient land was available to install the land-based pilot plant. There were also questions raised regarding the serious environmental degradation that could result from such an operation (Ravindran, 2005).

Nearly two decades ago, this was a much discussed method to become practical and widespread in any parts of the world. The sun heats the oceans and its energy gets stored in the topmost layers where the temperature rises to more than 25°C. Below the upper layers and at a depth of about 1000 m, the temperature drops to a few degrees above the freezing point. Such a difference in temperature is maintained permanently throughout the year from which energy can be obtained. The principle behind an OTEC plant is simple. A "working fluid" like ammonia or propane with a low boiling point is pumped into a closed tube exposed to the warm water zone. The heat of the warm water vaporizes the working fluid. This vapor is then taken to the colder water zone where it condenses to get back the working fluid in liquid form. If this vapor is allowed to expand through a turbine, it can turn into a generator. The vapor leaving the turbine is channeled into a condenser at a lower temperature of the cold water zone. The condensed liquid is pumped into a boiler-evaporator to start the cycle afresh. A diagrammatic representation of its working principle is given in Figure 8.2.

Most of the plants operate on what is called a closed-type OTEC. The designs available today are largely those of the anchored plants, either offshore or near the coastline. Floating types from ships have also been tried as pilot plants but without success. Modified versions of the OTEC include the entire heat exchangers placed on the shore. Only one long tube going up to 1000 m or more brings the cold water and exposes the shore-based

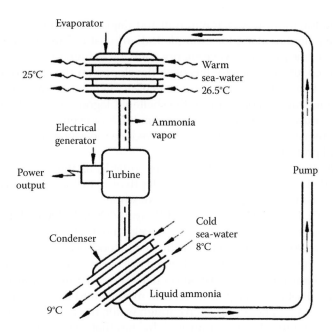

FIGURE 8.2
Concept behind an OTEC plant. A working fluid (e.g., liquid ammonia) is pumped through the heat exchangers (evaporator and condenser) and its junctions are regulated by the warm seawater (25–27°C) taken from the surface and cold sea-water (8°C) pumped from deeper layers.

condenser of the heat exchanger to get back the working fluid in liquid form. A very small tube pumps the warm sea-water from the surface to vaporize the liquid.

Realizing of the suitability of the Indian waters for OTEC, Tamil Nadu has recognized that OTEC could possibly be an ideal substitute for conventional energy generation. OTEC plant was planned to be set up off the Tamil Nadu coast, where four sites were considered for technical feasibility. The development schedule of an OTEC plant needs the study of a number of components to select the best site under all conditions. It needs precise determination of temperature and salinity profiles along many stretches the coastline in different seasons and a detailed knowledge of weather conditions.

8.4 Ocean Wave Energy

Ocean wave energy concept is based on the principle that the surface water of the ocean is kept in a state of continual motion. This continual motion can be harnessed to produce power. The vertical rise and fall of successive waves is used to activate either a water-operated or air-operated turbine. Another method of working the turbines uses the to-and-fro rolling motion of the waves. In a third method, waves are concentrated in a converging channel or basin and the breaking waves maintain a head of water whose concentrated momentum can drive a turbine.

In the Indian Ocean region, wave energy potential along the Indian coast is not as high as that obtained in the northern and southern latitudes. Wave height data collected near Kakinada indicated that wave heights are between 2.5 to 3 m during the nonmonsoon periods (Raju and Ravindran, 1985; Ravindran and Raju, 1997). This suggests that the average annual wave potential along for example Indian coast is 5 to 15 km. When calculated on this assumption, the potential wave energy could be as highest as 60,000 MW. The present level of research and technology has been developed on the basis of (i) multipurpose wave regulator system (WRS) incorporating electricity generation from absorption of wave energy from a long wave barrier and (ii) the sheltered lagoon on the shore side of the bar can be used as a natural harbor, a space for aquaculture, or even for coastal transport using light aircraft. Moreover, the barrier may as well reduce the action of wave erosion, thereby helping in coastal protection.

Wave regulator systems are considered more suitable, as they are positioned at a distance of about 500 m from the shore at a water depth of 10 m. Sites such as described above exist at a several locations along Tamil Nadu's coastline.

McCormick (2007), Falcao (2009), Khan and Bhuyan (2009), USDOE (2010) have identified 50 wave energy devices at different phases of development. The detailed dimensional scale magnitude of the wave devices is yet to be fully understood, because the wave devices in the direction of wave propagation is generally limited to lengths below the scale of the dominant wavelengths that characterize the wave power density spectrum at a particular site.

It appears that half of the energy is in vertical motion, and half as horizontal motion as can be seen in Figures 8.3 and 8.4. In deep enough water, the wave particle motion reduces exponentially with depth. The significance of this lies that a surface device can exploit the difference in particle motion between it and a deeply submerged object to extract power rather than having to be rigidly connected to the bottom of the ocean.

Along the Indian coasts, according to various reports and atlases published on waves, wave potential is not as high as in countries located in higher latitudes. The average wave

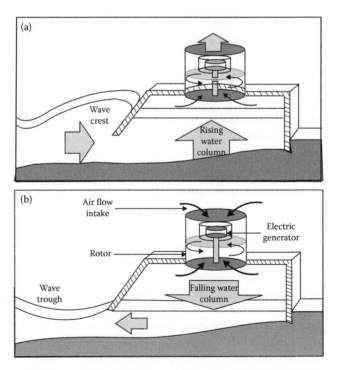

FIGURE 8.3
(a) Wave energy convertor (falling water column), and (b) wave energy convertor (rising water column).

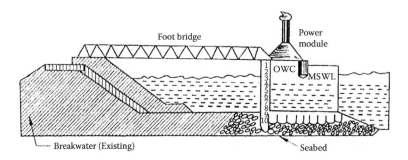

FIGURE 8.4
Cross-sectional plan of the wave energy device and its connection with the breakwater by a steel foot bridge.

power potential along the Indian coasts is between 5 and 10 kW/m. Scientific wave data using wave rider buoys for the Indian coasts are available from a few places only. Of these, the Kerala coast and the fishing harbor of Vizhinjam showed an average wave power potential of 13 kW/m length of the wave crest.

Research on wave energy commenced in 1983, with the formation of an interdisciplinary group at new Department of Ocean Engineering at the Indian Institute of Technology (IIT), Chennai, which has developed excellent facilities for conducting investigations on wave energy where both theoretical and experimental studies for the development of shape and optimum dimensions of the OWC device are being conducted (Figure 8.3). The fishing harbor at Vizhinjam, near Trivandrum, was selected, and the design and installation of a caisson and the construction of the power module was done by the same IIT group. Subsequently, improvements and refinements in the pilot plant including the modifications in the turbine and a greater efficiency of the plant were carried out by Prof. M. Ravindran and his colleagues. In fact, he and his colleagues are even now monitoring the performance of the pilot plant.

The pilot plant undoubtedly demonstrates how a random and diffused form of energy in the waves can be converted into electrical energy that can be exported via the electrical grid. In the existing wave energy plant, more than 80% cost goes to civil construction (concrete caisson). Considerable cost savings can be obtained using the concept of multifunctional breakwater wherein a power module could form an additional work incorporating a caisson into the breakwater. The breakwater could then provide the provision for more than one or a number of power modules.

The wave energy plant operating in India undoubtedly demonstrates a clean and renewable conversion process from waves to electricity. For India, it is undoubtedly a source of pride that we are in the forefront of this technology as its demonstration has been achieved largely using indigenous components. Although currently underutilized, ocean energy is mostly exploited by just a few technologies: wave, tidal, current energy, and ocean thermal energy.

8.5 Tidal Energy

Tidal energy may be developed, principally on the basis of a rotary hydroelectric generation plant. In China, such plants based on tidal energy are functional with an average installed capacity of 0.8 to 1.8 MW using the air pressure ring buoy is already well known for small-scale electricity generation from waves.

From the list of most feasible is producing energy from tides. Tides are much more stable than the other factors that are dependent on seasons. The powerful and regular ebb and flow of the tides are produced by astronomical gravitational forces of the sun and the moon. When the difference between the low-tide and the high-tide is large, and either natural or artificial mechanisms of water storage are available, tidal energy conversion is generated. The total global estimate of tidal power however is only 2% of the world's potential hydro-electric capacity. In the Indian Ocean, the most prospective locations for tidal energy are the Bays of Cambay and Kutch in India (Sharma, 1986; Sharma and Sharma, 2013; DART).

The principle involved in the generation of power from tides is easy to understand. The flow of water associated with the rise and fall in coastal areas including bays and estuaries gets advanced and retarded with the incoming and outgoing tides respectively. If a dam is constructed in such a way that a bay or basin gets separated from the sea and a difference in the water levels is obtained between the basin and the sea, power can be generated from the difference using low head turbines. Tidal power in the form of electricity is obtained from the oscillatory flow of water during filling and emptying of nearly closed coastal basin by the tides. There are a number of schemes that can be used for tidal power generation. The construction of a power house, sluice way, and dykes is essential for tidal power plants. The most important factor is the design and construction of a generating unit consisting of low-head turbines.

In addition, a low embankment about 4 m high would be necessary across to isolate the basin during high tide. It is proposed that the barrier would be connected to high ground level on either side to provide a permanent roadway over the barrage. Data that has so far been collected in the course of the study (wave current and tidal measurements as well as observations of bathymetry, sediment content and salinity) point to the suitability of the site for development. It seems the most likely problem would arise from the considerable level of sediment transport in the area. It has been suggested that the provision of the barrier across the Phang and Sang creeks with sluices could prevent this potential problem.

An estimate of potential ocean energy resources is given in Table 8.2. Of these, research on OTEC is the furthest advanced with the immediate possibility of establishing a pilot scale plant in the Laccadive, Andaman, and Nicobar Islands. There is little hope for commercial production in the near future and in all cases considerable technological development is necessary.

The Indian tidal energy sector is attracting overseas companies as well. Atlantis Resources, a UK-based tidal energy company, has secured a significant contract in an effort to utilize the power of the sea around India for the first time. The company has signed a deal with western Indian state of Gujarat, in order to establish the feasibility of developing tidal power projects that are capable of generating more than 100 MW of power, enough to supply to about 40,000 households (Tables 8.3 and 8.4).

TABLE 8.2

Potential Indian Tidal Power

Site	Mean Spring Range (MSR)	Mean Deep Range (NMR)	Average
Gulf of Kutch	6.43	4.02	5.23
Gulf of Cambay	8.75	4.79	6.77
Sunderban Ganges Delta	4.3	1.63	2.97

TABLE 8.3

Tidal Power Plants in India

Plant Owner	Plant Developer	Plant Type	Plant Location	State	Installed Capacity (MW)	Status
West Bengal Renewable Energy Development Agency (WBREDA)	National Hydro Power Corporation	Tidal Power Project	Delta of Ganga, Sundarbans	West Bengal	3.75	Announced
Gujarat Renewable Energy Development Agency	–	Tidal Power Project	Gulf of Kutch	Gujarat	900	No update available
National Institute of Ocean Technology	National Institute of Ocean Technology	Wave energy projects	Vizhinjam Fisheries Harbor	Kerala	0.15	Commissioned

8.5.1 Tidal Energy Site Selection and Fabrication of Plant

Indian Ocean, like other ocean and seas, can be used as a source of energy in the tropical region. This is possibly the most valuable source of energy for the island comments. To achieve this, however, specific technology development is needed before commercial-size plants can be established. At this time, it looks like the OTEC plant designed is not economically and commercially viable for small output of the order of 1 MW or less. Therefore, the

TABLE 8.4

Organizations Working on Ocean/Tidal Energy in India

Organization	Website	Activities	Additional Detail
MECON Ltd.	www.meconlimited.co.in	Preparation of status report on global wave energy for MNRE	–
West Bengal Renewable Energy Development Agency (WBREDA)	www.wbreda.org	Preparation of detailed project report (DPR) covering several aspects relating to the impact on physical, biological, and human aspects such as topography, hydrology, water and air quality, forest and vegetation, fauna, aquatic ecology, rehabilitation, services, health and education, etc. for setting up of a 3.6 MW tidal power plant at Durgaduani Creek in Sundarbans	Project cost: Rs. 40.15 crore
National Hydro Power Corporation	www.nhpcindia.com	Updation and upgradation of DPR (by WBREDA) and setting up of a 3.75 MW tidal power plant at Durgaduani Creek in Sundarbans	Project cost: Rs. 53.98 crore. Project completion period: 33 months
National Institute of Ocean Technology	www.niot.res.in	Wave and tidal energy demonstration plants in India	–

Source: Tidal Directory.

ocean wave energy appears to be more suitable and promising for island territories. Large OTEC plants of the order of 100 MW or an above shall be more economical and should be useful for mainland territories as well. Also tidal energy is useful in the coastal areas.

8.6 Sweet Water from the Sea: Low-Temperature Thermal Desalination Applications for Drinking Water

Sea water (volume $1.37 \times 10^9 \, km^3$) is a huge natural resource of salt, magnesium metal, and its compounds as well as bromine. The world's oceans contain about 5×10^{16} tonnes of mineral matter, of which 85.2% is sodium chloride (Mero, 1965). The earliest recorded extraction of salt from the sea appears to have been in China around 2200 BC (Schott, 1976). At present salt is recovered in India by traditional methods. In the period 1990–2000, the annual production of mineral from sea water was 169 million tonnes, of which salt constituted about 93%. Nearly 37% of all salt produced and 99% of all bromine comes from the oceans. The production centers of chemicals from sea water in India are based on quite large solar marine salt operation along the west and east coasts of India. Magnesium oxide has been produced from solar salt bitterns in Saurashtra and Maharashtra regions.

Uranium is dissolved in sea water in trace concentrations (0.003 mg/l) and the total in the world oceans in extraction is now thought as promising. Extraction experiments have already succeeded at laboratory levels, but there are many problems to be solved for commercial recovery. Present research projects are directed at the study of absorbent materials and development of the technological systems for uranium extraction from sea water including construction of a pilot plant. There has been some interest in uranium in organic-rich diatomaceous mud (as off Namibia), which has an uranium content in the range of 21 ppm (Schott, 1976) or in phosphatic nodules as off the Chatham High, southwest Pacific, which has a uranium content in the range of 7–435 ppm.

The Central Marine Chemicals Research Institute at Bhavanagar has fabricated a pilot plant for the production of sweet water from seawater. In addition, the institute is conducting research on the extraction of chemicals and metals from sea water.

NIOT has developed indigenous technology for utilizing the resources from the sea. It has developed a technology for generating freshwater from the sea water by using the temperature gradient of ocean called low-temperature thermal desalination (LTTD) process. In LTTD, the surface sea water from upper layers is evaporated in a low-pressure chamber and the resultant pure vapor is converted back to potable water in a condenser that makes use of cold sea water from the deeper layers of the ocean. Two important elements are the following:

1. *Island Desalination for the Sea Water:* Lakshadweep islands are remote and are facing drinking water scarcity due to increased population and tourism activities. A 100 m^3/day land-based plant was commissioned at Kavaratti Island in Lakshadweep in 2005. This plant has been continuously generating fresh water for the past 11 years to meet the drinking water needs of the island community. The water is of excellent quality. The plant is housed in a structure onshore. The bathymetry at the island is such that 10–12°C water is available at a depth of 350 m at a distance around 400–450 m from the shore, which is the source for cold water. The cold water is brought to the surface through a 600 m long high-density polyethylene

(HDPE pipe). The surface sea water is available at 28–30°C, which is the source for warm water. This first ever plant has become the main source of drinking water for the islanders and health of the people has improved considerably. This indigenized technology has been deployed in two more islands of Lakshadweep, namely Agatti and Minicoy, in 2009. LTTD plants in six more islands (Amini, Androth, Chetlat, Kadamat, Kalpeni, and Kiltan) in the Lakshadweep Island and lagoons are now on the anvil.

2. *Meeting Mainland Water Requirements from the Sea Water:* A barge-mounted desalination plant with a capacity of 1 MLD was installed successfully and demonstrated at 40 km off Chennai coast around 800 m water depth for mainland applications in 2007. The challenges include design, installation, and maintenance of plant in deep water and transport of product water to the coast. The single point mooring used for 1 MLD plant was the deepest in Indian waters. For the first time in the world, a 1 m diameter and 750 m long HDPE pipe was towed, upended, and connected to the bottom of a barge to pump deep-sea cold water at around 10°C. The 1 MLD offshore LTTD plant successfully generated freshwater for several weeks offshore. With the confidence gained, NIOT proposed a scaled up LTTD plant of 10 MLD capacity. Considering the large size of all the components and it being the first ever plant of its kind, which also has a major emphasis only on the offshore component, it seemed prudent to embark on this activity with industry partnership. With the help of industry, a detailed project report (DPR) for an offshore desalination plant of 10 MLD capacities is ready. In-house experimental and numerical studies on offshore components such as platform, moorings, interconnection components, and others were also carried out to augment the design of floating offshore plant in partnership with the industry.

8.6.1 Water Requirement of Coastal Power Plant

The LTTD technology was further utilized for the generation of freshwater from waste heat from coastal thermal power plants and a pilot project with capacity of 1.5 lakh liters per day was successfully demonstrated at the North Chennai Thermal Power Station (NCTPS) using their condenser reject water. Water was generated for long periods when all units were up and was supplied in the NCTPS premises. An experimental LTTD setup at NCTPS was installed in which experimental data was continuously logged and studied for performance evaluation of heat exchangers and demisters.

NIOT is attempting to scale up the LTTD technology at coastal thermal plants using an industrial partner. As part of it, it was proposed to install two LTTD modules each of 1 MLD capacity using power plant condenser reject at Tuticorin Thermal Power Station (TTPS), Tuticorin. Out of two modules, one module will be producing freshwater of quality less than 200 ppm and the other module will be producing boiler quality water. It was decided to work with industry and the design is now available. The plant can be installed using this design.

8.6.2 Self-Powered Desalination Plant at Islands

Efforts at NIOT are now focused on powering desalination using OTEC. Toward this aim, a laboratory setup to run OTEC and desalination cycle is being planned. A 2 kW OTEC turbine using R134a as a working fluid was designed for the first time ever in-house

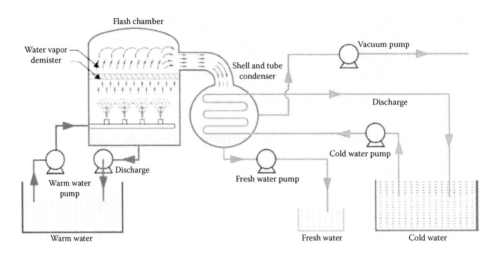

FIGURE 8.5
Diagram showing the principle of low-temperature thermal desalination. (Courtesy: NIOT.)

at NIOT. Other components such as heat exchangers, flash chamber, chillers, and others required for this laboratory were also designed and fabricated and currently are being readied for integration to set up the OTEC-Desal laboratory. Further, a proposal to establish an OTEC-powered desalination plant of 100 m³/day capacity at Kavaratti in the Union Territory Lakshadweep (identified as a priority program by the government of India) has been undertaken as a next logical step toward achieving self-powered desalination plant using OTEC. Toward this objective, various plant configurations were studied and the best configuration with single cold water pipe was finalized. Analysis of bathymetry data is completed and report is nearing completion. For this study, the design parameters have been finalized based on the surface and deep-sea temperature data (Figures 8.5 through 8.8).

FIGURE 8.6
Showing NIOT's first installed low-temperature desalination plant. (Courtesy: NIOT.)

FIGURE 8.7
Showing distant view of NIOT's plant. (Courtesy: NIOT.)

8.6.3 Benefit to the People of Lakshadweep

To the people of Kavaratti Island, the commissioning of this plant resulted in the availability of good drinking water, in which has led to a considerable improvement in their health. The NIOT has trained a local technician to maintain and operate the plant. This experiment has also benefited NIOT by generating self-confidence in its singular achievement. This is the first plant in the world to use the concept temperature difference in desalination. The whole concept was designed by the NIOT personnel. This success story is likely to be replicated in other islands of the Lakshadweep group.

FIGURE 8.8
Barge-mounted desalination plant of NIOT. (Courtesy: NIOT.)

References

Cohen, R., 2009. An overview of ocean thermal energy technology, potential market applications, and technical challenges. In: *Proceedings 2009 Offshore Technology Conference*, Houston, TX, USA, 4–7 May 2009.

DART, www.dartdorset.org/html/tidal.shtml.

Falcao, A., 2009. The development of wave energy utilization. In: 2008 Annual Report, A. Brito-Melo and G. Bhuyan (eds.), *International Energy Agency Implementing Agreement on Ocean Energy Systems*, Lisboa, Portugal, pp. 30–37.

Indian Wave Energy, National Institute of Ocean Technology, www.niot.res.in/m1/mm1.html.

Khan, J. and Bhuyan, G. S., 2009. Ocean Energy: Global Technology Development Status. Report prepared by Powertech Labs for the IEA-OES, Document T0104, Implementing Agreement for a Co-operative Programme on Ocean Energy Systems (OES-IA), Lisboa, Portugal.

Khan, J., Moshref, A., and Bhuyan, G., 2009. *A Generic Outline for Dynamic Modeling of Ocean Wave and Tidal Current Energy Conversion Systems*. Institute for Electrical and Electronic Engineering, Piscataway, NJ, USA, p. 6.

McCormick, M., 2007. *Ocean Wave Energy Conversion*, Dover Publications, Minola, NY. ISBN-13:978-0-486-46245-5.

Narin, A. E. M. and Stehli, F. G. (eds.). 1982. *The Ocean Basins and Margins*, Vol. 6. The Indian Ocean, Plenum, New York, pp. 1–50.

Raju, V. S. and Ravindran, M., 1985. Ocean energy in the Indian context, Mahasagar 18, No 2. National Institute of Oceanography, pp. 211–217.

Ravindran, M., 2005. Harnessing of the ocean thermal energy resource. In: H. K. Gupta (ed.). *Oceanology*. Universities Press, Hyderabad, pp. 26–28, 222.

Ravindran, M. and Raju, V. S., 1997. Wave energy: potential and programme in India. *Renewable Energy*, 10(2/3), pp. 339–45.

Sharma, H. R., 1986. Kachchh Tidal Power Project, Water for Energy. *3rd International Symposium on Wave, Tidal, OTEC and Small Scale Hydro Energy*, Brighton, England, pp. 179–189.

Sharma, R. C. and Sharma, N. 2013. Energy from the ocean and scope of its utilization in India. *Int. J. Env. Eng Management*, 4(4), 397–404.

USDOE, 2010. *Energy Efficiency and Renewable Energy Marine and Hydrokinetic Database*. Energy Efficiency and Renewable Energy, US Department of Energy, Washington, DC, USA.

Vega Luis, A., 1992. Economic of ocean thermal energy conversion. *Ocean Energy Recovery*, 152–181.

9

Remotely Operated Vehicle and Autonomous Underwater Vehicles

9.1 General

One of the new technologies associated with the recent increase in underwater activity is the Remotely operated vehicle (ROV) and underwater autonomous vehicles (AUV). The scientific and economic significance of underwater technology is considerable, particularly as the ocean is becoming more and more important not only as the source of energy, mineral material, and food but also due to its influence on climate change. Therefore, underwater technology particularly at great depths offers a very special challenge for AUVs (Hornfeld, 2005). ROV and AUV technology is utilized in underwater operations, such as in the offshore hydrocarbon operations. This reduces the expense of conventional manned submersible systems. It has replaced humans in the performance of subsea intervention tasks. Here we examine the growth and future of ROVs and AUVs in subsea exploration and mining. A comparative analysis of manned and unmanned underwater systems is described along with advantages and limitations of both the systems. A description of the past and current ROV and AUV application is presented with an account of the major uses of ROVs and AUVs within the oil and gas industry. Projections are made on the future oil and gas needs and the types of ROVs and AUVs that may meet these requirements and likely to be used in deep-sea mining (Busby, 1978).

Recent advances in technology are identified together with current trends in the development of ROV and AUV systems. Attention is given to ROV and AUV systems employed by the offshore hydrocarbon recovery. It deals briefly with the operation of ROV and AUV and conventional diver systems and details the development of AUV or free swimming vehicles (FSVs) and problems associated with the development of this type of vehicle are outlined. The development of dedicated ROV and AUV support vessels and the range of ROV and AUV launch and retrieval mechanisms are described and assessment made of their relative advantages and shortcomings.

It also enumerates ROV and AUV uses outside of the offshore oil and gas industry. The use of ROV and AUVs by the power and telecommunications industries is examined, relevant ROV and AUV types described, and future potentials assessed. Developments in the use of ROV and AUVs in marine mining projects are investigated and innovative projects for future ROV- and AUV-based mining programs described, including an innovative nodule collection system. The application of ROV and AUVs to fisheries and fisheries science is assessed and potential areas for ROV and AUV employment investigated. Environmental monitoring by ROV- and AUV-based systems is discussed and recent developments in this field are outlined. Finally a range of innovative potential ROV and AUV uses are considered. The role of the naval force in ROV and AUV development is examined. The major tasks and applications of ROV and AUV systems in naval defence usage are identified, with particular attention being paid to mine countermeasures and the role of autonomous vehicles. Future naval defence markets and potential improvements are investigated (Busby, 1986).

9.2 Development Perspective

Significant increase in underwater activity associated with the offshore hydrocarbon industry led to emergent technology associated with the increase in the ROVs and AUVs. Presently ROV and AUV technology has come to dominate underwater operations, largely at the expense of conventional manned submersibles and has begun to replace humans in the performance of underwater tasks.

ROV and AUV development has been closely linked to hydrocarbon exploration and exploitation. Initially the majority of subsea intervention tasks were performed by divers, operating from the surface or from lock-out submersibles, or by manned submersibles.

ROV and AUVs are by definition unmanned; however, some systems feature dual role capability and can operate in manned or remote modes. A complete ROV and AUV system generally consists of the vehicle itself, in some cases a supporting garage or launcher, and an umbilical cable, which links the vehicle or its garage to a surface (shipboard) control and display console such as Canadian Ropos. Shipboard support equipment often includes a launch and retrieval device of some description (A-Frame, crane) and a cable winch, which may incorporate a tether management system and vessel heave eliminator. Also present will be an enclosed area for the vehicle operator (or "pilot") and the shipboard maintenance equipment. Should shipboard power supplies prove inadequate or unavailable, a power supply unit would also be required.

There exist several types of remotely controlled underwater work systems that have been designed and constructed for a range of operational tasks (Figure 9.1). Systems in operation include towed vehicles, used in deep sea and seabed survey and exploration, and seabed crawler-type vehicles employed mainly in pipeline and cable laying and burial. The largest section of the ROV and AUV market services the offshore oil and gas industry and is composed almost entirely of self-propelled, free-swimming cable controlled (tethered) vehicles.

An ROV and AUV is essentially a platform-bearing propulsion, navigation, and sensory (usually TV camera) equipment. To this basic design may be added a range of specialized sensory, tool, and manipulator packages. ROV and AUV designs are flexible and varied and encompass a range of vehicles, from totally free-swimming systems to structure-based systems (Graham, 1985; Jaeger and Wernlie, 1984).

The vehicle generally consists of an open metal framework, which serves to protect and support the various components. Syntactic foam blocks generally provide vehicle buoyancy, which is usually slightly positive by a few kilograms and is offset by ballast, adjusted according to the required tool and sensor payload. Buoyancy is top mounted as an aid to overall vehicle stability. Smaller ROV and AUVs are often totally encased in a resistant shell, which offers a low hydrodynamic profile, useful when vehicle mass is low and power supply limited. Vehicle weights range from around 13 kg (Sea Technology Inc.'s SEA SCANNER) to 405 tons (Ferrostaal AG's SUPRA); however, most vehicles have a mass of between 1000 and

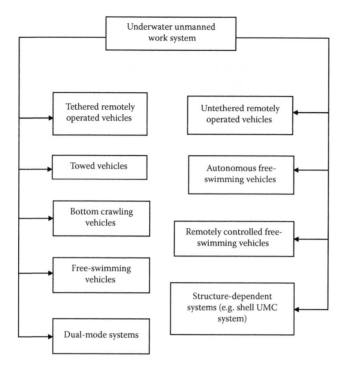

FIGURE 9.1
Breakdown of underwater unmanned work systems and indicates those systems forming bulk of ROV and AUV.

2000 kg. Many of the larger vehicles are trenchers and pipeline and cable burial systems. These are generally "one-off" designs and are operated on a lease basis.

Recently it has been found useful to classify ROV and AUV systems into five categories, based on vehicle size and task capability. Classes used are

1. 100-Class observation/inspection vehicles (e.g. Mini-ROV and AUVER)
2. 200-Class observation/light work vehicles (e.g. SPRINT, RCV-225)
3. 300-Class medium work vehicles (e.g. SCORPIO, RIGWORKER)
4. 400-Class heavy work vehicles (e.g. SOLO, DRAGONFLY)
5. 500-Class heavy/specialist vehicles (e.g. Brown and Root Mechanical Underwater Trencher)

Such a classification is purely descriptive and no firm distinction between classes exists.

9.3 Tasks and Applications

Since the majority of ROVs and AUVs are employed within the offshore oil and gas sector, most tasks and applications are necessarily related to hydrocarbon exploitation and exploration. Nonoil- and gas-related ROV and AUV usage will be considered therein (Kerr and Dinn, 1984).

A number of ROV and AUV tasks are commonly identified:

- *Diver monitoring/observation:* Tasks include observations of diver safety and general welfare, confirmation of worksite, and provision of lighting facilities. "Eyeball"-type ROVs and AUVs can provide immediate task assessment by surface-based experts who may then offer firsthand advice to the diver. Some ROVs and AUVs also carry diver-operated tools and may assist in tool retrieval.

- *General structure inspection:* Regulations require annual or biennial structural inspection of offshore platforms to assess damage and extent of marine fouling. Checks are also made on debris and scour. Such tasks require general visual surveys and observation.

- *Detailed structure inspection:* These tasks are carried out on a five-year cycle in order to meet platform recertification requirements. Tasks include simple measurement of corrosion protection (CP) systems and detailed inspection of nodes using advanced nondestructive testing (NOT) techniques.

- *Pipeline inspection:* This was one of the first tasks to be effectively performed by ROVs and AUVs and is now almost totally carried out by ROV and AUV systems in preference to divers. The requirement is for a vehicle carrying a large payload of sensors integrated with the systems aboard the surface support vessel. Accurate position fixing and navigation are imperative and a high-speed capacity and station holding capability may also be necessary.

- *Site and route survey:* Requirements include a stable platform carrying a range of instrumentation, which often includes side-scan sonar, sub-bottom profilers, and cameras. Coring equipment may also be required and a good navigation system is essential. The system should also be capable of relatively high speeds, up to 9 knots, to allow high speed "live boating."

- *Drilling support:* ROVs and AUVs are particularly preferred for drilling support tasks in deep water conditions. Manipulator dexterity, accurate position fixing, and large payload capability are required together with the ability to mount a range of tool and sensor packages, video suites, and short-range sonar systems. Good station-holding capability in high current speeds may also be demanded.

- *Construction support:* A broad heading, which includes structural repair and installation. General tasks include measurement, cutting and joining, and line or cable attachment. Operational requirements call for a range of tool packages and manipulator dexterity plus station maintenance capability.

- *Debris removal:* Tasks range from retrieval of dropped items to the attachment of lifting cables to larger objects, which may pose a nuisance or hazard in the immediate vicinity of underwater structures.

- *Structure cleaning:* Usually involves the removal of marine growth and fouling organisms. Generally good station maintenance is provided by a range of suction cups and grabbers. Cleaning packages include abrasive pads and brushes, and high-power water jetters.

- *Pipeline and cable burial/repair:* Usually undertaken by 500 class ROVs and AUVs, which belong either to the bottom crawling or larger free swimming designs. Burial tasks demand a trencher and back filler, together with a cable handling facility. Requirements are for powerful vehicles, possibly combined with large mass. Precise positioning is essential.

- *Special purpose systems:* A rather amorphous category including vehicles built for specific locations and tasks. An example is found on the Shell underwater Manifold Centre (UMC), which is maintained by a vehicle mounted on a set of rails surrounding the UMC.

The ultimate goal of ROV and AUV development is the replacement of humans underwater entirely, or the reduction, to the lowest possible level of any risks involved with diving tasks. Moreover, diver operations are often of short duration, human-intensive and, hence, expensive in terms of time and money, extremely so when periods of operational downtime are considered. Diving bottom time (i.e., when the diver is actually on site and working) can range from 70% to as low as 30% of the total port to port contract time. Furthermore, dive preparation time alone contributes over 10% to the on-site operation duration. The majority of manned submersibles have already proven themselves commercially unattractive in most underwater tasks (Marsland and Wiemer, 1985). ROVs and AUVs possess several advantages over conventional underwater systems, particularly in the performance of simple tasks such as inspection and observation. With improvements in technology, they should prove capable of replacing divers and manned submersibles in tasks of increasing complexity; indeed the increasing role of ROVs and AUVs in the performance of underwater tasks has been much at the expense of manned submersible systems.

Several major potential advantages of ROVs and AUVs may be identified (Holmes and Dunbar, 1975)

1. ROVs and AUVs possess lower capital costs than manned submersibles of equivalent capability.
2. ROV and AUV operational costs are generally 10%–50% lower than those of conventional diver operated systems (Marsland and Wiemer, 1985).
3. For the majority of underwater tasks, and for the purpose of all current oil and gas industrial depth requirements, ROVs and AUVs have no depth limitations. In this respect, they score considerably over conventional diving methods in terms of safety, time, and expense considerations in deep-water operations.
4. Despite representing a large capital investment (compared to a diver), ROVs and AUVs are, in an emergency situation, an expendable asset. In an emergency, there are no time constraints upon rescue and in the event of total systems failure, they may be hauled to the surface using the umbilical cable.
5. Diver operations suffer limitations relating to air supply duration, cold tolerance, and fatigue. Despite providing a more comfortable working environment, manned submersibles are also constrained by power supply and crew fatigue. ROV and AUV operations benefit from a comfortable working environment and operator fatigue can be avoided by shift working. Thus ROV and AUV operations may proceed uninterrupted for 24 hours a day.
6. The provision of surface image displays permits simultaneous access to real-time information, rather than just the onsite diver/submersible crew member.
7. The smaller mass of most ROVs and AUVs provides greater maneuverability and within structure access than that available to manned submersibles. Furthermore, suction arms and mechanical grabbers permit better station maintenance in high current speeds than that experienced by divers.

8. Small- to medium-sized ROVs and AUVs may be used from any ship of convenience, which possesses an adequate derrick. Very small ROV and AUV systems may be used simply "over-the-side."

9. The umbilical tether links the vehicle to a virtually unlimited energy supply, thus eliminating the need for batteries and thereby reducing weight.

10. In some cases, the use of low-level light intensity and SIT cameras often provide better visibility than that available to an onsite diver.

Currently available ROV and AUV systems do exhibit several deficiencies, especially when compared to the "man-on-the-spot" situation. However, these are viewed by many ROV and AUV manufacturers, not as drawbacks, but as potential areas for improvement. Such improvements are viewed as the main future prospects for the ROV and AUV industry. The limitations of AUV/ROV and AUVs over conventional underwater systems are the following:

1. Unsatisfactory imaging techniques. Faults lie in a lack of visual resolution, depth perception, and perspective and peripheral view. This is mainly due to a reliance on conventional 2-D television imagery. Operator opinion is geared toward a requirement for a windscreen' view.

2. Inadequate dexterity. Manipulators are not, as yet, able to provide a replacement for the human hand. Many manipulators suffer from a lack of "feel" or force feedback mechanisms experienced by the remote operator. This disadvantage is now some way to being overcome.

3. Reliance on an umbilical cable for power and signal transmission causes problems associated with drag, weight, and entanglement. A vicious circle is experienced where increased vehicle power requirements necessitate an increase in cable diameter, which leads to greater drag and weight. Cable enlargement often puts severe limitations on within structure worksite access and increases operator stress.

4. Smaller ROVs and AUVs often experience difficulties with station maintenance in high current speeds.

5. Operational tasks within the splash zone, such as scour inspection and marine fouling removal lead to specific problems associated with support ship and vehicle instability.

6. Shortcomings exist with a range of tool packages. The problem is particularly prevalent among detailed inspection packages for NOT and magnetic particle inspection (MPI).

7. High-sea states may cause problems during launch and retrieval unless advanced vehicle handling facilities are available. Difficulties may also be encountered with ship motion and noise, leading to failure of the vehicles acoustic navigation system.

The application market for ROV and AUV equipment can be broken down into three major sectors:

1. Exploration/survey
2. Construction/support
3. Inspection, maintenance, repair (IMR)

9.4 Developmental Trends in ROV and AUV Technology

ROV and AUV development has been largely determined by two major factors: specific task requirements and improvements in underwater systems technology.

The primary impetus to ROV and AUV evolution has been supplied by the demands of the offshore oil industry. The twin aims of replacing humans underwater and in providing the cost effectiveness of underwater operations have seen a transformation in ROVs and AUVs from the simple "flying eyeball" systems used in observation and inspection tasks to a more diverse array of systems capable of complex intervention tasks.

Recent technological advances are numerous but may be categorized as falling into the following general areas:

- *Cameras/Optics:* Recent advances have included low-light level and SIT cameras, which are often fitted as standard on current ROVs and AUVs. The development of stereoimaging and advanced photogrammetric techniques are still in their formative stages.

- *Positioning:* Positioning and navigation systems must be accurate and reliable, consistent and able to interface between surface and subsea systems. Recent advances include improvements in side scan sonar and continuously updated sonar displays. Accurate acoustic navigation is of paramount importance to efficient ROV and AUV operation and new developments in supershort and ultrashort baseline tracking reflect this requirement. Acoustics, interfaced with surface positioning systems, now provide directly potable ROV and AUV locations. Features such as gyro stabilization and preset height control are incorporated into many models as standard. Obstacle avoidance packages are in the early stages of development.

- *Umbilical cables:* Cable drag, weight, and a tendency to snag are major limitations to ROV and AUV operations and reduce within-structure maneuverability. The introduction of fiber optic technology has provided a method of substantially reducing overall cable diameter, while increasing the rate and scope of two-way signal transmission. Cables are generally of low diameter but are capable of handling high voltage power, low-voltage data, and optical signals in one integrated package. Reductions in cable weight and drag permit the extension of ROV and AUV operations into deeper water.

- *Tether management systems:* An alternative answer to the problems associated with cables has been the incorporation of a garage or deployment cage, either as standard equipment or as an option. The deployment cage is lowered by cable from the surface support ship to a position alongside the intended work site. The ROV and AUV then emerges from the cage, connected to it by a short operating tether. This system eliminates many of the problems experienced with a direct vehicle-to-surface cable, such as difficulties with surface vessel heave and cable entanglement.

- *Manipulators:* Recent developments in manipulators have been aimed at the production of a system capable of mimicking the action of the human hand. A number of ROVs and AUVs now incorporate manipulators with up to seven degrees of freedom. Systems incorporating touch control and force feedback are also available.

- *Tool packages:* Several task-specific tool packages have been developed for a range of underwater work requirements. These include water jetters, wire brushes, and flooded member detectors, CP probes, thickness gauges, and cable cutters. One-off structure-integrated tools may also be designed and incorporated according to customer requirements. Experience with nodal cleaning packages indicates that ROV and AUVs mounting manipulator-held tools are not efficient and the answer to this problem probably lies with dedicated automatic work packages with an ROV and AUV acting as a tractor.

- *Electrical, mechanical, and hydraulic systems:* Early ROV and AUV models were commonly prone to thruster failures, hydraulic problems, water leaks, and electrical faults. Developments in systems design have led to much greater system reliability. Cable connectors are now greatly improved and less prone to leakage while manipulators are generally operated by seawater hydraulic systems, thus avoiding the problem of seawater ingress experienced with oil-based hydraulic systems. The overall designs aim at ease of maintenance, further aiding system reliability.

- *Electrical safety:* External power supplies in ROV and AUV systems may reach 3000 volts. Therefore efficient electrical grounding is essential, particularly if divers are in attendance. Recent safety recommendations and guidelines put forward by the Association of Offshore Diving Contractors (AODC) have stimulated the development of ground fault monitoring systems. These are intended to continually monitor internal electrical resistance and in the event of an electrical leakage will shut down the entire system within 25 milliseconds (well within the safety threshold).

9.4.1 Trends in ROV and AUV Systems

Overall trends in ROV and AUV development can be briefly summarized as a move toward smaller, cheaper systems, on the one hand, and larger, deeper capability systems on the other.

About one hundred ROV and AUV models are currently available. Within this range of designs, emphasis appears to be on the production of three major types to fit the requirements of the offshore oil and gas industry (Daniel, 1986):

- Basic, heavy duty work vehicles, designed to accept a variety of modular work and tooling packages and capable of accepting a large payload, generally of the Class 400 type (e.g. GEC/OSELs DRAGONFLY). These are widely accepted as representing the most efficient system for deep subsea servicing.

- Systems designed to undertake a specific task or range of tasks, generally platform cleaning/inspection. These generally fall within the Class 300 category. Increased specialization has led to the development of structure-dependent, task-specific systems.

- Low-cost (LC) ROVs and AUVs. There has been a strong demand for an inexpensive ROV and AUV system, capable of efficiently performing visual inspection/observation tasks and, possibly, some light manipulative work. LC ROVs and AUVs have, in some part, been developed in order to offer an inexpensive alternative to larger ROV and AUV systems in the performance of these tasks. Another attribute of LC ROV and AUV systems is their ease of deployment and "user-friendly" operating requirements. Flexibility is a major contributory factor to their success, many vehicles being capable of accepting sonar, CP, NOT, and high-quality stills and video camera equipment, along with acoustic navigation systems. An improvement in positioning technology has also greatly contributed to the success of this class of ROVs and AUVs (Smith, 1985).

9.4.2 Autonomous Underwater Vehicles

AUVs, also called free swimming vehicles (FSVs), may operate in one of two general modes: (i) unsupervised (fully autonomous, preprogrammed) and (ii) supervised (acoustic- or radio-controlled). The advantages offered by the development of such a vehicle are numerous; for one thing, all cable-associated problems are eliminated. However, signal transmission through sea water is a problem, difficulties being encountered with signal attenuation and acoustic multipath (secondary signals) effects. Signals transmission must be via acoustic links, radio frequencies (except extremely long wave) having a limited range. ISE of Canada do produce a semisubmersible, radio-controlled FSV (DOLPHIN) for shallow water and coastal survey.

One solution to signal transmission problems is the use of a deployment cage system as outlined previously, but with an acoustic link between the vehicle and garage. This system has the advantage that signals travel horizontally and do not traverse several layers of water and, hence, do not suffer from thermocline- and halocline-induced interference. For longer ranging vehicles, some sort of surface acoustic link would be required. Real-time data links are a major problem to overcome; acoustic signals transmission is slow and a 30 frames per minute TV data update is the best possible. Thus several seconds may elapse before obstacle avoidance can be carried out. If the FSV is travelling faster, such action may be too late. Research into acoustically transmitted video imaging is currently underway.

The answer to all the above difficulties could lie in the development of a fully autonomous vehicle. Such a vehicle would have an artificial intelligence with a facility for obstacle avoidance and would be preprogrammed to find its way to a work site, carry out specific tasks, and then return. Several research programs into autonomous vehicles are underway, but future markets await developments in artificial intelligence systems.

9.4.3 Support Vessels

Initially ROV sand AUVs were operated from vessels of convenience. Essentially, any ship able to provide deck-space for the ROV and AUV and its ancillary equipment were considered suitable. Such vessels included supply boats and diver support vessels and even trawlers. It remained the ROV and AUV contractor's responsibility to provide facilities such as a steady power supply, workshops, and handling gear. The entire ROV and AUV system was housed in a container that could be easily conveyed directly to the ship's deck. It is now being recognized that major improvements in operating efficiency and cost effectiveness may be achieved through long-term installation of ROV and AUV systems on a dedicated support vessel. One of the greatest advantages offered by such a vessel is the increased deployment capability, which can permit operations in extreme sea state conditions.

A ROV and AUV support vessel should feature

- A capacity for dynamic positioning, permitting good station maintenance
- Thruster redundancy
- Multiple ROV and AUV capacity, to increase operations capability

- Ample deck space and adequate work area
- Surface and subsurface positioning systems giving integrated accurate positioning of vessel and ROV and AUV
- Adequate accommodation for operators and support personnel
- Moon pool launch facility

Additional facilities can include computerized data processing, photo processing laboratory, and video inspection room.

9.4.4 Launching Gear

In recent years, improvements have been made in the launch and retrieval equipment used with ROV and AUV systems. Drawbacks identified with the use of simple unmodified cranes and winches included the problem of vessel heave and wave oscillation, particularly in high-sea states. Advanced tether management systems have been developed that incorporate electronic sensors, which monitor load, pitch, and roll and adjust the cable loading accordingly via microprocessor-modulated controls. For example, the ROV and AUV deployment system produced by Scomagg Hydraulics Ltd. offers active heave compensation and can recover ROVs and AUVs without snatch in force 10 conditions. Some problems may still be encountered with wave impact when the ROV and AUV are entering the water. A range of options are available for launching and retrieving subsea systems: over the side, stern-handling, and open-handling, each having its advantages and drawbacks. A further option is the moon-pool approach, essentially a hole through the deck and hull of the ship. This offers protection from wave action reduces roll and pitch of the vehicle and costs relatively little to install in an existing ship when compared to the construction of a specialist vessel.

9.5 Nonhydrocarbon-related ROV and AUV: Uses and Technology Transfer

A small number of ROV and AUV manufacturers have met with success in marketing ROV and AUV systems outside the oil and gas sector with the marine hydrocarbon sector currently at a standstill, or at best making slow progress with new developments, there has been a slight move toward market diversification by both ROV and AUV manufacturers and service companies. One major exception to this trend has been the military market where there has been a demand for ROV and AUV systems since their inception. This section will examine the alternative uses for ROV and AUVs outside of the oil and gas sector.

9.5.1 Mine Countermeasures

By far the largest single use of ROV and AUVs by the military is for mine inspection and neutralization, or mine countermeasures (MCM). This is one area where the removal of man to a remote position is highly desirable. Leaders in this field, at present, are ECA of France. Sales of this system are now approaching the 300 mark. Two years later, most commercial ROV and AUV companies could offer alternatives to the system.

9.6 Future Trends

It has been realized that the future for ROV and AUVs will lie in the design and manufacture of specialized vehicles equipped for specific tasks such as diver support, pipeline inspection, NOT inspection, and drilling support (Smith, 1985). This view was stimulated by the discovery that many of the intermediate-size ROVs and AUVs available were underpowered for NOT and inspection tasks. The manufacturer's solution to this is a larger vehicle able to operate power cleaning and inspection equipment, featuring an inspection arm capable of entry into areas of limited access.

A number of underwater tasks offer smaller markets, for which specialized designs are not a commercially viable proposition. It is in this area that modular designs will make more of an impact, tool packages being interchanged to meet the specific task requirements. Problems are still encountered with umbilical snagging, and tether management systems will continue to improve (Walsh, 1985; Newman, 1986).

Operational capability will benefit from a move to make control consoles and displays more "user-friendly." In this area, it is expected that the recent trend in the use of modern information display techniques drawn from technology designed for the aerospace industry will continue. Certain areas of systems design and instrumentation will benefit from further research and development. These include

1. Improvements in acoustic positioning technology and obstacle avoidance
2. Improved capacity for station maintenance, with better access to confined work sites and improved within-structure management packages
3. Development of autonomous systems, particularly through the development of independent power sources (including battery development), preprogramming techniques, and artificial intelligence
4. Improved underwater communications technology, possibly by lightweight optical fiber cables or by lasers

There is a huge potential for low-cost ROVs and AUVs in inspection and general-purpose tasks, where they will be required to perform light work tasks and mine neutralization. Further improvements in manipulators, level of sophistication, and artificial intelligence will all serve to make commercial systems more attractive to the military. Portability is another important factor when considering military systems where rapid deployment is essential. Commercial ROV and AUV systems could acquire more appeal to military buyers through the development of more "user-friendly" control systems (Hampson and Bergen, 1986).

At present, the technology developed for the offshore oil and gas sector is, due to the oil price collapse, relatively inexpensive. The situation offers an excellent opportunity for the military to obtain several commercial systems for advanced field trials. Moreover, military research into ROV and AUV systems is now a difficult and expensive proposition in view of the extensive range of technology systems available: modified commercial ROV and AUVs offer a proven technology combined with shorter procurement periods. One drawback, however, of present acoustic controlled systems is that acoustic signals are not secure. A solution may be found in the development of laser-based communication systems.

9.6.1 Cable Laying

Advances in commercial ROV and AUV technology have not been exclusively in the off-shore oil and gas market. The last 10 years have seen the development of a variety of underwater vehicle systems employed by the power and telecommunications industries. Such ROV and AUVs perform tasks involving cable burial, location and retrieval, and trenching. Detailed topographical surveys are performed on proposed cable routes; however, fine control of cable laying operations is necessary and vehicles must be equipped with cameras and sonars for obstacle detection and should be capable of obstacle avoidance (Nield, 1986).

9.6.2 Deep-Sea Mining Operations

The excavation and removal of material from the seabed, at increasing water depths, has gained in importance, primarily in conjunction with oil and gas exploration activities.

Continuation of subsea structures, repair of damaged pipes and cables, and the digging of "glory holes" to protect subsea installations from iceberg scour all require the excavation of materials. These requirements have already led to the development of diver-operated systems, which are able to contend with a variety of seabed materials. This material is extremely variable, ranging from sand and gravel, through silt, mud and clays, to drill cuttings and drilling muds containing rock fragments.

The Alluvial Mining Company (AMC), in response to request for a ROV- and AUV-operated excavator, has developed a system intended for use with the CUTLET ROV and AUV. The main purpose of this system is the salvage of embedded equipment and wreckage (Hill, 1986). AMC has also developed the Alluvial Mining Remotely Operated Dredging System (AMROD). This is capable of static use or may be carried on a ROV and AUV. It can operate down to depths of 300 m and has successfully excavated a variety of materials. Although used only in oil and gas projects, it has considerable potential for the fast removal of overburden and the selective mining of a variety of mineral deposits.

A MITI project in Japan is investigating the use of a system incorporating a conventional hydraulic lifting device with a towed, possibly tracked, collecting vehicle. Similarly, France is highly dependent on overseas territorial deposits of key minerals, particularly nickel. This dependence has led to an interest in the mining of deep-ocean strategic mineral deposits, which was realized in the formation of a nodule mining program in the early 1970s. The scope of the project has now been broadened to include a greater range of deep-ocean seabed uses. Future vehicles will also be tethered in order to increase power availability and telemetry capability. Deep-sea and continental-shelf surveys of mineral resources is well within the capability of the present generation of ROVs and AUVs: full exploitation of oceanic mineral resources only awaits the development of extraction technology, concomitant with favorable economics and the resolution of any legal disputes.

9.6.3 Fisheries

There are several examples where it has proven economically feasible to transfer technology, developed primarily for military and industrial uses, to fisheries applications. Examples include microprocessors and sector scanning sonar, which now find use in hydroacoustic fish detection and enumeration devices, the so-called "fish finders."

The oil industry and the military have already underwritten the developmental costs of ROV and AUV technology and it is now available for use by fisheries scientists and the fishing community (Thorne, 1986).

Potential fisheries applications of ROV and AUVs include

1. Use in studies of fishing gear characteristics. ROVs and AUVs can be used effectively both around stationary capture devices (e.g., crab pots) and on moving gear (e.g., trawl nets). Using video techniques, the capture efficiency of any type of gear could be assessed. Furthermore, the impact of bottom trawling gear on benthic communities, particularly of alternative commercial stocks, could also be investigated. ROV- and AUV-based studies would prove more rapid than those done by divers and more relevant than laboratory-based studies on gear effectiveness.

2. Direct observations of fish stocks. ROVs and AUVs equipped with video cameras would prove particularly useful in rapid assessments of benthic stocks of shellfish, crabs, shrimp, and demersal fish. This sort of study currently relies on slow, diver-based investigations. ROVs and AUVs also permit precise position location.

3. As vehicles for acoustic sensing devices. The advantages of ROVs and AUVs in this area lie in the avoidance of boundary layer interference effects, either by using upward looking modes or by close proximity to the seabed using a downward looking mode. Available ship-borne sonar cannot accurately determine fish concentrations found on, or close to, the seafloor. Acoustic sensing devices score over rival techniques as fish cannot detect and do not react to sonar as they do to lights.

4. General behavioral studies. For example, studies of fish schooling behavior, predator avoidance behavior, and response to fishing gear. Despite a large amount of research in this area, little real information has been gained due to the reliance on laboratory based experiments or surface observation through clear water.

5. Aquaculture assistance. Activities in this area could include the monitoring of the condition of sea cages, repair of cages, and general observational and measurements tasks.

With the current depressed state of the oil-based commercial ROV and AUV market, the availability of inexpensive ROV and AUV technology for fisheries applications could lead to an increased usage of ROV and AUV systems, including the use of low-cost ROV and AUV systems for tasks such as net inspection and net or propeller disentanglement.

9.6.4 Environmental Monitoring

Increased government and industrial concern and responsibility for the impact of human activities on the marine environment have led to an increase in the number of environmental surveys and monitoring programs. Generally, environmental surveys involve intensive sampling of a statistically valid number of precisely located, often widely dispersed stations. Until recently, such a survey would have necessitated a large number of short-duration diver excursions. Suitably equipped ROVs and AUVs provide an ideal alternative in this type of task, particularly for work in deeper waters. Use of ROVs and AUVs permits visual site selection and precise station positioning. ROVs and AUVs also allow rapid-visual surveys of the seabed, which may be analyzed by a number of experts and can provide valuable information on environmental conditions.

Potential sensor developments include pipeline leak detectors (through the use of introduced fluorescent dyes), in situ hydrocarbon level measurement devices, and radiation counters. Furthermore, on site collection of water and sediment and organic samples for subsequent laboratory analysis could also provide useful information on environmental status. The potential of ROVs and AUVs as monitoring tools is, at present, virtually untapped.

9.6.5 Other Uses of ROVs and AUVs

The early military vehicles, usually modified commercial systems, were used for general-purpose tasks. These tasks still include inspection, wreck location, site and route survey, search and retrieval, submarine rescue, diver assistance, torpedo and weapons system recovery, hull inspection, and harbor route conditioning.

Some examples of the use of military-owned ROVs and AUVs are the inspection of the stranded Soviet submarine off Karlskrona by Swedish Navy SEAOWLS and the location of downed aircraft (notably the Air India 747 and the Korean Airlines Flight 007) by US Navy ROVs and AUVs. It is significant that, in the recovery of the Air India 747 flight recorder, the US SCARAB system was used as no UK-produced system was capable of reaching 2000 m. Six US Navy ROV and AUV systems were also employed during the search and recovery of the *Challenger* shuttle remains.

Further, alternative uses of ROVs and AUVs cover essentially all underwater tasks that have hitherto required human presence. Inspection and observational tasks could and do provide a wide market, particularly for low-cost ROVs and AUVs. In Canada, ROVs and AUVs find frequent employment for dam and inland waterway inspections, particularly in tunnels. Other areas for inspection include power plant cooling water intakes and outlets, sewage outfall pipes, and nuclear reactor water pools. One recent innovative use has been the internal inspection of water ballast filled petroleum storage tanks.

Civilian law enforcement agencies and rescue services could provide another source of employment for inspection and manipulative capability ROVs and AUVs.

Ocean science and engineering research would benefit immensely from ROV and AUV technology, both for purposes of pure research and for educational purposes. The current low-cost of ROV and AUVs puts them within reach of most institutional budgets. Increased ROV and AUV usage by academia could also generate spinoffs of value to the ROV and AUV industry in return. Recreational use of ROVs and AUVs is a possibility, which cannot be dismissed. Indeed, Sea Scan Technology includes recreational purposes in the sales literature for their SEASCANNE & LCROV and AUV system. Future uses of ROV and AUV systems seem unlikely to be limited by price considerations but rather by the limit of the imagination of any operators whose work requires the performance of underwater tasks. The demands of the offshore oil and gas industry have been the primary formative influence on the evolution of ROV and AUV technology. Government and academic research interests have contributed to ROV and AUV development to a lesser extent. Academic research has so far generated a small number of spinoffs with high commercial potential.

ROV and AUV development has been stimulated by safety considerations and economic factors. ROV and AUV systems are now used either in conjunction with conventional diving techniques or have replaced divers entirely in the performance of a large number of underwater intervention tasks. This trend appears likely to continue, concomitant upon technological advances in ROV and AUV systems, particularly in deep-water operations.

ROVs and AUVs are presently capable of performing a multiplicity of subsea tasks within the marine hydrocarbon sector. Several classes of vehicle may be distinguished, according to vehicle size and capability. ROVs and AUVs exhibit several advantages over conventional manned systems and the disadvantages they do exhibit are viewed as technological challenges offering future developmental prospects.

Recent advances in ROV and AUV technology have largely centered on cameras and optical systems, acoustic positioning techniques, umbilical cable design (use of optical fibers) and tether management systems, manipulator dexterity, and tool package versatility. The key to the recent success of ROVs and AUVs can be attributed largely to improvements in acoustic positioning. General systems reliability and safety considerations have also recently undergone improvement. Trends in ROV and AUV development follow a move toward smaller, cheaper systems or toward larger, deeper capability systems. The requirements of the oil and gas market are served by three major ROV and AUV types: basic heavy duty work vehicles of modular design, systems designed for the performance of specific tasks and low-cost remotely operated vehicles (LCROV and AUVs), performing inspection, and some lightwork tasks. The latter type of ROV and AUV offers the greatest potential for ROV and AUV use both within and outside of the hydrocarbon industry. A range of ROVs and AUVs have also been developed specifically for diver support duties. Currently research is progressing into tetherless, AUVs. Problems have so far been encountered with acoustic telemetry links, particularly in the transmission of real-time video images. Truly autonomous vehicle development is awaiting progress in artificially intelligent systems research.

The use of dedicated ROV and AUV support vessels and improvement in ROV and AUV handling systems have considerably enhanced the efficiency of ROV and AUV operations, particularly in adverse weather conditions and high-sea states. ROV and AUV operational efficiency could be further enhanced by designing subsea systems for ROV and AUV intervention, rather than, for example, developing manipulators as analogues for the human hand.

Future ROV and AUV requirements are expected to demand deeper water capability and increased systems redundancy and reliability. ROV and AUV operations are expected to extend into harsher conditions and to tasks such as the maintenance of subsea production complexes (Partridge, 1984). The ROV and AUV market, having initially undergone an exponential growth period, now appears to be almost static. This is primarily due to its ultimate dependence upon oil prices and the resulting oil company inactivity. The dependence on world oil prices confers a degree of fragility upon the ROV and AUV industry, which could be avoided by product and market diversification. One area in which ongoing requirements for ROV and AUVs do exist is in the performance of inspection, maintenance, and repair (IMR) tasks. This outlet may reduce against the adverse effects of the current oil price slump. ROV and AUVs find further use in the power and telecommunications industries, and offer potential for use in deep-sea mining projects. Additional potential uses exist in fisheries science, environmental monitoring, and a range of underwater intervention tasks. Simple inspection tasks probably offer the greatest potential use of ROVs and AUVs in the short term.

India possesses a beginning in making ROV and AUV industry, occupying three levels: manufacturer, component producer, and leasing agent. Despite its present attachment to the offshore hydrocarbon industry are growing application in deep-sea structure and mining, provided some diversification of product and target market is undertaken and fuller use made of the potential of ROV and AUV systems in a wider variety of uses, the future of the Indian industry can be viewed with optimism.

References

Busby, F., 1986. Undersea vehicles—the military side. *Sea Technol.*, 27(1): 19–24.

Busby, R. F., 1978. Engineering aspects of manned and remotely controlled vehicles, *Phil. Trans. R. Soc. Lond., A.*, 290: 135–152.

Daniel, J., 1986. *Trends in ROV and AUVs, in ROV and AUV Review.* Windate Enterprises Inc., California, pp. 125–127.

Graham, D., 1985. Low cost vehicles dominate ROV and AUV show, *Sea Technol.*, 26(5): 28–40.

Hampson, D.J. and Van Bergen, G., 1986. A navel approach to object classification for military requirements—Present and Future. *Proceeding ROV86 Conference*, Aberdeen, Graham and Trotman Ltd., London, pp. 327–339.

Hill, J. C. C., 1986. Dredging tools for ROV's. In: Remotely Operated Vehicles— Technology Requirements, Present and Future. *Proceeding ROV86 Conference*, Aberdeen, Graham and Trotman Ltd., London, pp. 164–177.

Holmes, R. T. and Dunbar, R. M., 1975. Seabed surveying by remote control, *Oceanol. Int.*, 1975: 75–80.

Hornfeld, W., 2005. Proceedings of OMAE2005. *24ND International Conference on Offshore mechanics and Arctic Engineering*, Halkidiki, Greece, June 12–17, 2005, pp. 10.

Jaeger, J. E. and Wernlie, R. L., 1984. ROV and AUV'84 report: Technology update from an International perspective, *Mar. Technol. Soc. J.*, 18(1): 69–77.

Kerr, A. J. and Dinn, D. F., 1984. The use of robots in hydrography. *Proceedings of Oceanology International Exhibition and Conference*, Brighton, Society for Underwater Technology, p. 9.

Marsland, G. E. and Wiemer, K., 1985. DAVID: A Versatile Multipurpose Submersible Support System for Remote Control or Diver Assisted Performance, in Submersible Technology, Advances in Underwater Technology, Ocean Science and Offshore Engineering, Vol.5, Graham and Trotman Ltd., London, pp. 407–412.

Newman, J., 1986. *The Commercial ROV and AUV Business: New Equipment Sales, in ROV and AUV Review*, Vol. 2, Windate Enterprises Inc., California, pp. 4–5.

Nield, E., 1986. Safe Routes for Undersea Cables, *New Scientist*, No. 1509, pp. 38–41.

Partridge, D. W., 1984. Future developments of ROV and AUVs, *Underwater Technol.*, 10(1), Spring Issue: 1–17.

Smith, J. D., 1985. Are ROV and AUVs efficient? In: *Submersible Technology, Advances in underwater Technology, Ocean Science and Offshore Engineering*, Vol. 5. Graham and Trotman Ltd., London, pp. 339–406.

Thorne, R. E., 1986. Some Applications of ROV and AUVs in Fisheries Science. In Remotely Operated Vehicles: Technology Requirements—Present and Future, *Proceedings of ROV and AUV'86 Conference*, Aberdeen 1986, Graham and Trotman Ltd., London, pp. 354–359.

Walsh, D., 1985. The Low-Cost Remotely Operated Vehicles (LCROV and AUV), in Submersible Technology, Advances in Underwater Technology. *Ocean Science and Offshore Engineering*, 5: 343–351.

10

Seabed Mineral Exploration and Deep-Sea Mining Technology

10.1 General

Mineral deposits of economic significance mostly occur in small areas, which is due to different ore-forming processes involved. The idea that "it is usually easier to locate a mine by chance than to predict one" is no longer to be accepted, thanks largely to the methodology and our better understanding of the principles of ore formation. However, the immensity of the oceanic areas involved and the very high cost of research therein are strong arguments for a strategy: a concept for prospecting the seafloor, at least to predict areas without any chance of economic mineral concentration. Much of our thoughts have also been modified during the last four decades when marine geophysicists and geologists have developed the concept of seafloor spreading and plate tectonics. It is related to an increased heat flow from mantle of the earth combined with cracks and faults produced and continuous upcoming of the earth's crust beneath the continents and oceans. The thus generated hot mantle material penetrates the surface, forming basaltic lava. Both sides of these moving apart from the "spreading centers" and push into each other's zone. We have seen that the rate of spreading varies more than 10 cm/year. A continuous supply of basalt fills the gap and forms a new oceanic crust. These processes are at present concentrated near the axial lines of most of our oceans, resulting in upcoming of midoceanic ridges generally with water depths of 2000–3500 m. The spreading oceanic crust cools and subsides. Therefore, the water depth increases landward when it reaches about 4000–5000 m, when a critical situation arises for the carbonate particles produced by marine organism. They completely dissolve in the water depths. Normal midoceanic sedimentation rates of some cm per 1000 years are then reduced to some mm per 1000 years. Such very low sedimentation rate favors the growth and the survival of manganese nodules on the ocean floor.

At present, two major types of seafloor metalliferous deposits, products of distinctly different processes, are the focus of world attention. As we have seen in Chapters 2 and 3, manganese nodules and crusts, which cover large areas on the seafloor of the major ocean basins, are formed by sedimentary, concretionary, and biogenic processes. The metals they contain may be from hydrothermal or sedimentary sources or be concentrated by the geochemical reaction of seawater and sediments.

The second types of major metallic deposits are more exciting because one can observe their formation. Further, it is observed that in the fracture zones, faults, and the spreading centers of the seafloor, hydrothermal springs discharge solution containing iron, manganese, cobalt, zinc, copper, silver, lead, gold, and other metals. A few of these metals have precipitated in the form of carbonate, oxide, sulfide, sulfate, or silicate minerals in crusts and chimneys or stocks around hydrothermal vents. Some are disseminated in the sediments and siliceous oozes or mud in the seafloor, forming metalliferous sediments. Details of their occurrence have been discussed in earlier chapters (see Chapter 4) (Figures 10.1 and 10.2).

10.2 Exploration Methods for Manganese Nodules

Exploration of manganese nodules on the seabed is an expensive but challenging exercise. A well-equipped research ship is a prerequisite for any such program. Various techniques involving sampling device have been developed to bring material from water depths of 3000 to 5000 m. Conventional dredges have been redesigned for optimum collection. A dredge is sunk to the ocean bottom and slowly dragged for 1–2 km on the seafloor. However, for more accurate and quick surveys, a new system of sampling has been developed, which is called "free fall grab" sampling device. This device is now increasingly used for its easy handling and accurate sampling (Figure 10.1). An advantage in the search for manganese nodules is that fact that they lie on the top of the sediment and are readily visible (Figure 10.2). Therefore, information of underwater topography is essential. This is done through new developments in echo-sounding and underwater television with deep-sea cameras and video-recordings suitable for exploring the population, extent, and distribution of deposits. For the progressive determination of the amount of valuable metals, samples are brought to ship or laboratory and analysed using any of the known techniques, such as x-ray fluorescence, energy dispersion isotope technique, and others.

Exploration campaigns are taken up after due diligence and detailed work plans. This will involve consideration of several aspects, which have been summarized in Figure 10.2. One has to depend on both indirect and direct methods to observe the geology and nature of the deposit, and one may refer to several relevant books (Kunzendorf, 1986; Roonwal, 1986). For manganese nodule, detailed exploration is generally conducted by a system of sampling called free-fall-grab sampling. A resource evaluation is subsequently carried out, on the basis of average weighed grade of total Cu + Ni + Co sum as 2.27%, which is considered economical.

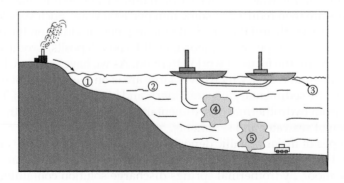

FIGURE 10.1
The concept of nodule mining (1—Ocean slope, 2—Main ship, 3—Mother ship, 4—Intermediate state, 5—Module collection).

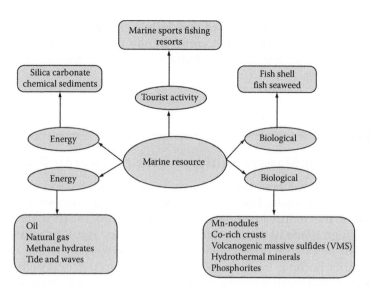

FIGURE 10.2
Showing the different seabed resources.

Mining, cost, and economics of Mn-nodules would need to be moved out. The eventual mining of the nodules will also need several questions settled. Because the ore is spread on seafloor, the equipment must be able to strip large areas. For this purpose, several systems have been experimentally tested, such as bucket arrangements, remotely controlled tractors on the seafloor, dragging baskets with opening of various sizes separating unwanted pieces, and a kind of vacuum cleaner device with several rotating suction heads, to mention only a few. Likewise ore beneficiation and metallurgy techniques are also being worked out. The assumed metallic content of nodules in a prospective location may give per tonne (dry) nodule metal values. Using estimates of operation, transport to processing site, and shore handling, the cost of procuring the nodules is likely to be about $50 per tonne. This is based on assumed composition of 30% Mn, 1.5% Ni, 1.2% Cu, and 0.3% Co.

Nodules of grade high enough to mine have been reported mainly from six areas identified earlier. Among them, the two regions, namely South Central Indian Basin (10°–25°S: 70–85°E) and South Australian Basin appear most promising. This assessment is based on the percentage of stations reporting manganese nodule occurrences in these areas. The average weighted grade as 1.2% Ni plus 1.03% Cu (total Cu + Ni 2.27%) is considered economical. Since nickel is a rarer metal compared to copper and is 220% costlier than copper, the economic considerations of manganese nodules are often calculated on this basis. This would mean that Cu + Ni in nodules has been defined as (Cu + 2.2 Ni) ≥3.76%. Similarly, Cu + 2.2 Ni must total at least 3.0% for cutoff grade. According to this yardstick, the Central Indian Basin emerges as a possible site for first-generation seabed mining. Nodules here are high in Ni + Cu and their average grade is similar to grade from the North Central Pacific zones. But the available data indicate that nodule abundance is low. Therefore, further exploration in this area was conducted for an assessment to reveal valuable mine sites. The metal values found in these areas range upto 25% Mn, 1.5% Ni, and 1.5% Cu. These deposits are at or below the CCD and are underlain by siliceous ooze. The region in which they occur is thought to underlie an area of high biological productivity in the surface waters, as do similar deposits in the Pacific, and the deposits may prove to be analogous to their Pacific counterparts. Survey work conducted by the NIO-DOD/MoES has recognized and fully evaluated the economic potential of the deposits in the two similar areas Central Indian Basin. This is the basin in which these two areas have been registered by India with the International Seabed Authority (ISA) for future mining.

10.3 Exploration for Massive Sulfides on the Midoceanic Ridges

Marine exploration for minerals during the past few decades has led to discovery of several marine mineral deposits. The most exciting of these discoveries are marine sulfide deposits found on the midoceanic ridges and subduction zones. They are generally localized and show high grade of metal content in them. They result to active undersea volcanism, generally at a depth of 2000–3000 m. Their discovery in 1979 is one of the most significant scientific findings in marine science. The regions that host such massive sulfide deposits have provided earth scientists with excellent opportunity to understand primary ore-forming hydrothermal systems in real time. In addition to ore formation, these systems support a unique and varied fauna. These fauna depend upon this volcanic activity and metal nutrients for their existence. The quantity of minerals added to seawater by this volcanism has greatly helped in understanding the concept on the factors that contribute to the chemical composition of the oceans (Table 10.1).

Seafloor massive sulfides (SMS) comprise a range of minerals. Many samples collected during different exploration cruises have shown that the high metal content in them would make these deposits an extremely valuable resource. This has resulted in thinking that SMS may provide an alternative source of metals to traditional land-based supplies. However, at present not enough is known of the true abundance of the deposits, nor what represents an "average" deposit. But the results obtained to date have been encouraging. As more such high-grade deposits are discovered, the economic and commercial interest in SMS is expected to grow. It is therefore imperative to consider what technological methods and engineering aspects exist or are likely to be developed for eventual commercial mining to begin. Deep-sea mining is no longer confined to the drawing board. Many

TABLE 10.1

Exploration Strategy of Sulfide Mineral Deposit

Range of Sulfide Mineral (m)	Method
10^4 to 10^6	Regional sediment sampling (concentration of Fe–Mn)
10^4 to 10^6	• Regional water sampling applicable to actively accumulating deposits—measure weak acid-soluble amorphous manpower suspended particulate matter (ferric-hydroxoide) and δHe • Total dissolved manganese • Methane (CH_4)
10^3	• Bathymetry
10^3	• Magnetic, gravity, long-range side-scan sonar
10^2	• Short-range side scale sonar
10^1	• Bottom image, near-bottom water sampling
10^0	• Dredging, submersible, drilling, electrical methods

Source: Based on Rona, P.A., 1984. *Earth Sci. Rev.*, 20: 1–104.

methods and techniques have been developed keeping in mind to extracting manganese nodules at depths of 4000–6000 m, and a project in the Red Sea has proved the viability of pumping metalliferous mud from over a 2000 m depth. The systems enumerated here could be used for SMS mining. It needs to be understood that unlike the Red Sea's metal-rich mud and manganese nodules, SMS are hard consolidated deposits, similar in form to land-based sulfide ore deposits. It is important that any system for mining SMS need to include a mechanism for disaggregating the deposit prior to transport to the surface. At present, the technology for achieving this in the deep ocean is being developed by groups interested in them. Alternatives to disintegration have been suggested, such as solution mining or vent capping, but a physical crushing of the ore is viewed as the likely choice by those working on the best resolution of the mining process.

Technological requirements are not the only criteria that will dictate whether or not SMS ever become a viable economic resource. Consideration must also be given to nontechnological factors such as world demand for the contained metals, the status of deep-sea deposits in international law, and environmental aspects. Some details on the occurrence of SMS deposits on the Mid-Indian Ocean Ridge have been given in Chapter 4.

10.3.1 The East Pacific Rise and Other Promising Areas

The East Pacific Rise is an actively spreading ridge bisecting the seafloor of the southeast Pacific terminating at the west coast of North America to form the Gulf of California. It has provided some of the most exciting MPS finds to date, including the first observation of hydrothermal sulfides, forming chimneys around an emitting vent. A total of 25 active vents were observed emitting metal-laden fluids at temperatures of up to 350°C. In 1981 a deposit was discovered at 20°S on the EPR. It was located on a section of the ridge with a spreading rate of 16–18 cm/year (one of the fastest yet investigated). This deposit has yet to be fully delineated, but the intense volcanism and hydrothermal activity associated with such rapid separation make it appear a promising area for sulfide deposits of ore size. During GEOMETEP 4 expedition, extensive survey and sampling work was conducted along the EPR between 17°S and 27°S in December/January 1985/1986. The results were very encouraging (Marchig, Gundlach, and Shipboard party).

In 1982, a French expedition (CYATHERM) located a large SMS deposit on the EPR at 12°50′N, containing around 80 sites of active or relict venting. Chimneys were observed to grow at an astonishing rate (up to 8 cm/day) (Hekinian and Fouquet, 1985) and the largest deposit found was estimated to be 800 × 200 m. Originally this deposit was believed to be only a meter or so thick, but subsequent work by Francis (1985) using resistivity measurements has shown that it could possibly be up to 9 m thick in parts containing many thousands of tonnes of potential ore-grade material. Again this estimate is based on an extrapolation of very limited data and so should be treated with caution. This deposit is also interesting because it was found to contain relatively high levels of cobalt, which could be a significant factor in assessing its economic potential, due to cobalt's strategic value.

In the Atlantic Ocean, the Mid-Atlantic Ridge (MAR) system bisects the Atlantic Ocean. It has a very slow spreading rate (less than 4 cm/yr). Even though the MAR provided evidence for hydrothermal activity as early as 1973, no significant mineralization was discovered until 1985. It had been assumed by many that the MAR was too inactive to produce smokers and MPS mineralization, but in July 1985, active vents and hydrothermal deposits were discovered at the TAG field, 26°N, ironically very close to the site of the original discovery of mineralization 12 years previously. Rona has argued that theoretically, slow-spreading centers such as the MAR provide greater potential for the formation of mineral deposits large enough to warrant commercial investigation (17). If this be the case, then it is important that more exploratory work be conducted on slow spreading centers to assess the distribution and frequency of such deposits should they exist.

Another extremely interesting possibility is that of finding large SMS deposits in back-arc basins, which are also believed to be active regions of seafloor spreading where new oceanic crust is formed. Sillitoe (1982) suggested such areas as likely sites for the formation of SMS and recent discoveries appear to substantiate this hypothesis. Hydrothermal material was located by a German expedition in the Lau Basin and North Fiji Basin in late 1984, the same year the Japanese began investigations in the Okinawa Back-Arc Trough, which were to culminate in the discovery of an active hydrothermal mound. In December 1985, an Australian team located hydrothermal chimneys and associated fauna in the Manus Back-Arc Basin, Papua New Guinea, and finally, following discovery of hydrothermal plumes on the Mariana back-arc spreading center, sulfides and vents were discovered there in April 1987. How typical these finds are of back-arc basins will only be shown by more widespread and detailed investigation. The Mid-Indian Ocean Ridge, a slow-spreading system, has attracted attention most recently (Chapter 4).

The seafloor in many areas is peppered with undersea volcanoes, which are not directly related to plate margins of any description. Many are the product of "hotspot" activity. Each of these has the potential to host some form of sulfide mineralization and indeed many have been found to do so. Off-axis volcanoes provide conditions that are favorable for the formation of much larger SMS deposits than normally found on a ridge axis (Hekinian and Fouquet, 1985). Despite this, little detailed work has yet been carried out on seamounts.

10.4 Extent and Grade of SMS Deposits

It is apparent from the preceding sections that a reliable estimate of the global resource potential of SMS deposits is fraught with difficulty. Apart from the obvious paucity of useful data on the size and distribution of the deposits, there are technological difficulties in the collection of the necessary information (see Chapter 4). Before discussing the possible size of such a resource, it is important to define what is meant by SMS prospect. It is likely that sulfide formation at spreading centers is routine. Then it follows that all of the ocean floor will be peppered with deposits or their remnants. These will have been covered by subsequent lava flows and/or sediments. Are these also to be considered as a possible resource? At present no geophysical method exists, which could easily pinpoint such subsurface deposits, though it is conceivable that a suitable system could be developed. Even so, given the technical problems involved in merely dredging the seafloor in the deep sea, it is most unlikely that these relict SMS will be considered as an economic resource in the foreseeable future. So the following section will relate only to deposits on or near the seabed.

It is important that deposits which manifest themselves at the surface should be considered in three dimensions. It will be shown in Chapter 4 that systems have already been conceived which in theory at least, could mine a deposit down to its full vertical extent. This brings us to the crucial point of the probable extent of SMS deposits in the third dimension. Unfortunately, spreading ridges present special technical problems, which until very recently have prevented conventional drilling. Sedimentary cover is necessary to "key in" a traditional drill bit, and this does not normally exist on the new basalt found at spreading centers. Consequently, knowledge of the vertical extent of the deposits is almost exclusively inferred. Embryonic systems do exist, which will be able to estimate the thickness of the deposits using geophysical and electrical methods, and the Ocean Drilling Project has attempted to drill into a ridge crest this year using a special yoke to guide the drill bit. Work on land-based deposits, thought to represent ancient SMS analogs uplifted onto the continents (ophiolites), suggests that some form of stock-work mineralization may extend for 100 m or more below the surface in favorable geological conditions. However it would be premature to assume that the same can be applied to modern-day deposits until tangible evidence is obtained through drilling and coring.

The composition and grades of the deposits found to date have been extremely variable. Table 4.2 gives a fairly characteristic range of ore samples from SMS deposits compared with typical values of samples from land-based sulfide deposits in Cyprus and Canada. It can be seen that typical SMS grades compare extremely favorably with commercial grades currently being exploited on land. This is shown even more strikingly in Table 4.3. which attributes value per tonne to a range of sulfide ores, samples from Cyprus, and manganese nodules. Of course, such comparisons do not take into account the huge development costs required to exploit marine sulfides and the fact that any marine operation would have far greater running costs than the equivalent on land. Another feature of SMS highlighted by Table 4.3 is the wide range of values associated with different deposits. Indeed ores from the same deposit can show equally wide variations (Table 4.4), necessitating large-scale sampling to produce a viable assay. Estimates of the likely global potential of superficial, massive SMS deposits also vary widely. Cann (1980) points out that if we assume that an ore deposit of 3 million tonnes of ore containing 2% copper occurs every 100 km on the world's spreading ridges, this would give a total resource of 30 million tonnes of copper. This would provide a useful input into world supplies, but considered in the light of current consumption of around 7–8 million tonnes/yr, this is not a huge resource. Consideration of other metals from the ore would of course improve the economics Also such estimates do not take account of possible deposits on subduction zones or in back-arc basins, and more importantly may underestimate the extent of the deposits in the vertical dimension.

10.5 Technology Needs for Exploration and Mining

This chapter reviews the current, developing, and required technologies for the exploitation of MPS deposits. It focuses on the requirements for the exploitation of SMS finds located at spreading centers or seamounts, which manifest themselves on the seafloor. Buried deposits off-axis, should they be considered as a resource, would constitute a "second generation" operation, as the technical problems of location, proving, and extraction would be far more formidable. The sections follow the standard chronological order in the winning of any mineral deposit: location, proving, extraction and processing.

Early finds of sulfide material on the seabed were due in no small part to chance, but as knowledge of their genesis has increased, so has the marine geologist's ability to locate individual deposits. As in any geological exploration, deposits are located by a process of progressively "homing-in" from regional survey to pinpoint location. This process is well documented by Rona and Scott (1993), and the various stages of this activity have been summarized. The first stage in the search for sulfide bodies is the consultation of available bathymetric, topographic, and geological charts, which have been compiled by a variety of relatively mature technologies such as side-scan sonar, seismic profiling, and subsea photography. Having defined an area of general interest, of say 103–106 km² a more detailed survey of sediment and water composition can be employed to locate "hotspots" of volcanic activity.

Mapping of anomalous metal content in seabed sediments is now an established technique. Concentration of metals such as iron and manganese (Fe, Mn) are of particular interest. These concentrations are higher closer to the vent, after allowing for the effects of general oceanic circulation. Sampling of the lower half of the water column can also reveal strong indications of hydrothermal activity. Again, high levels of Fe and Mn in suspended particles provide evidence of hydrothermal activity, as does dissolved and particulate manganese measured as total dissolvable manganese (TDM). Guidelines for hydrothermal mineral deposits are given by Rona (1984).

The next stage of a program involves detailed mapping and delineation of the chosen site (say 100 sq. km). The first stage consists of traversing the area with a deep-tow vehicle. This comprises a platform with a variety of scientific instruments towed behind the ship at about 5–10 m above the seafloor. This could house sensors for all or a selection of the following: Fe, Mo, and He and CH_4: The package also normally includes a conductivity, temperature, depth (CTD) device, which obtains water samples via a rosette sampler for a detailed measurement of the contained elements, and a combination of sonar, side-scan sonar, and magnetometer to perform small-scale mapping of the topography of the seabed. Finally, video and high-resolution deep-sea cameras plus an array of temperature sensors allow real-time detection of emitting vents. Precise location, to within a few meters of the position of the vehicle at any given time, is achieved by the deployment of a net of three or more acoustic transponders on the seafloor.

Several suitable deep-tow systems are now available including Deep Tow (Scripps Institute of Oceanography) and SAR (IFREMER), and the US Navy's Argo/Jason system operated by Woods Hole Oceanographic Institute (WHOI), which located the wreck of the *Titanic*. The Argo system will have several innovative features including the tethered submersible Jason able to detach itself from Argo and maneuvre precisely on or around the seafloor for more detailed inspection of interesting features. This system is also being utilized to develop a concept known as "tele-presence," whereby stereoscopic TV cameras relay live pictures to the mother ship, giving the sensation of actually being on the seafloor at depths approaching 6000 m. Eventually it should be possible to relay such live pictures to participants thousands of miles away in land-based facilities, allowing "tele-conferencing" in real time.

These are sufficient to reveal areas of active volcanism within the area; relict sites are far more difficult to detect but may well be revealed by the TV cameras or magnetometer. There is a need for a method of producing large-scale optical images of the seafloor which can reveal the presence of relict deposits. It is also necessary to retrieve selective samples to estimate the value of a given deposit; two methods are possible: dredging and utilization of a submersible.

Simple dredging techniques such as grab samplers, claw and bucket dredges, box samplers, and piston core/vibra-core samplers may be employed in the early stages of the survey to give an idea of the overall nature of the geology. These will only give samples of superficial deposits and/or unconsolidated sediments. They also suffer from the disadvantage that the precise location of the sampling point is uncertain. Germany has partly overcome this problem by developing a grab sampler specifically for evaluating SMS deposits. The latest version has proved highly successful for the recovery of large site-specific samples of sulfide ore. The grab utilizes six grab segments to sample an area of over 4 sq. m and has a powerful cutting action, which can be increased by repeated opening and closing of the jaws (Figure 10.4). The closing action and in-built high-resolution TV cameras are self-powered,

FIGURE 10.3
TV-controlled pneumatic graph unit very useful for exploration for SMS. (Courtesy: Praussag, Germany.)

but the grab is guided to its target by the movement of the mothership. Capable of operating in depths of up to 10,000 m for 8 hours at a time, the grab represents the most powerful and cost-effective sampling tool yet developed for SMS (Figure 10.3).

Submersibles, manned or unmanned, provide an even more site-specific method for the detailed investigation of individual deposits, but their use is time-consuming and expensive (particularly in the case of manned submersibles). Table 10.2 lists the manned submersibles that have played a major role in SMS research. Of these, the most widely used have been Alvin, Cyana, and Pisces IV. Each one is capable of performing a range of tasks, enabling detailed sampling of the sulfides, vent emissions, and/or associated fauna. An obvious advantage is that they provide the opportunity to place a scientist at the site of investigation, allowing real-time decisions on the appropriate course of action. It is now under development, which attempts a compromise between these two extremes: tethered and autonomous. The first system is Argo/Jason from WHOI, which will combine the benefits of a tethered multipurpose ROV (Argo), with the maneuverability of a free-swimming vehicle (Jason), which is only connected to Argo by a light neutrally buoyant umbilical. At present, Jason is only equipped with high-resolution TV cameras, but eventually it will be provided with a wide range of manipulative and investigative devices. Another approach is that adopted by a group at the Marine Physical Laboratory (MPL) at Scrips Institution of Oceanography (SIO), USA, who have developed a towed vehicle, RUM III, which is also capable of crawling along the seabed like a small tractor. It has facilities for adjusting its buoyancy so that it can be towed in neutrally buoyant mode, or provides a stable platform on the seabed for drilling in a negatively buoyant configuration. Another possibility being investigated by MPL is to equip a "standard" multisensor towed vehicle with directional thrusters to enable it to operate in free-swimming mode near the seafloor. Eventually it may be possible to develop autonomous remotely operated vehicles with equivalent capabilities to the towed vehicles, but at present, major problems remain regarding power/data storage and telemetry.

Although the above developments show great promise, it is likely to be several years before they are totally operational. In the interim period, manned submersibles will

TABLE 10.2

Manned Submersible Used in SMS Exploration

Name	Persons	Water Depth (m)	Owner
1. Pisces IV	3	2000	Canadian–Russia
2. Shinkai 2000	3	2000	JAMSTEC Japan
3. Cyana	3	3000	IFREMER France
4. Alvin	3	4000	US Navy
5. Nautilus	3	6000	IFREMER France
6. Sea Cliff	3	6000	US Navy
7. Chinese	3	Details not available	China

continue to play a crucial role in the final small-scale investigation and sampling of SMS deposits. The latest generation of these is exemplified by Nautile, a state-of-the-art submersible developed by IFREMER in conjunction with the French Navy. Nautile carries three people to a maximum depth of 6000 m, giving it the capability to reach 97% of the ocean's seafloor (hence its original name SM-97). Its low weight (18.5 tonnes) is achieved by employing titanium alloy for the pressure vessel and external components, and low-density syntactic foam for buoyancy. It is equipped with a gripping arm and a manipulator arm, possessing five and seven degrees of freedom respectively. The Japanese organization JAMSTEC is also developing a three-man submersible capable of operating down to 6500 m, which has been successfully used in the Indian Ocean.

The formidable array of technological devices described above is only capable of analyzing a deposit in two dimensions. Submersibles can drill a core a few centimeters. deep, however, for the purpose of ore-body delineation, information regarding the vertical extent of the deposit is essential. Deeper drilling is the obvious answer, but this has proved extremely difficult on ridge crests.

Although drilling and coring in several thousand meters of water is now a relatively straightforward procedure, existing systems rely on sedimentary cover to "key-in" or spud the drill bit. Such cover does not exist on new oceanic ridges. This problem has been addressed as part of the Ocean Drilling project (ODP), an 14-year international program to drill thousands of sites on the seabed for the first time. A "bare rock spud" was designed and developed, which would rest on the seabed and allow studding of the drill bit and subsequent reentry. The base is lowered to the seabed and the orientation of the cone signaled to the mothership. If it is in a suitable position, cement is pumped into the box to stabilize it. A special hard rock drill assembly is then guided into the cone by acoustic methods and drilling commences (Hammett, 1985). This system was recently deployed in the southwest Indian Ocean, adjacent to the Atlantis II Fracture Zone, as part of Leg 118 of the ODP. Unfortunately, this attempt was unsuccessful due to the fragmented nature of the basalt, but a later attempt off-axis succeeded in drilling 500 m into gabbro in 750 m of water. It is unusual to find gabbro so near to the seabed and the coarse crystalline nature of the rock facilitates drilling. The effectiveness of this system for drilling into fresh ocean ridge basalts has not therefore been proven; however, other attempts are planned for the future on the EPR (13°N) and the Jaun de Fuca Ridge.

An alternative to full-scale deep sea drilling is a hard rock portable drilling system. Portable drilling systems have been employed for many years on the continental shelf in depths down to 500 m to obtain cores of sediments and hard rock. Various systems are available, developed in the United States, Japan, Canada, and the UK. Adaptation of this technology to operate on hard rock in the deep sea would be highly advantageous, portable drills designed for deployment from a standard oceanic research vessel would obviate the need for expensive dedicated drill ships. Attempts to achieve this transition have been made recently in Canada, Japan, and the UK, but two basic technical problems regarding power supply and stability remain. In many ways, the problem is similar to that for ROVs; solving the power problem by utilizing an umbilical adversely affects the stability of the drill. A further problem is that the umbilical cable must be left slack to allow drilling, which can lead to kinking thus damaging the wire and endangering recovery. Finally, the extremely rugged topography associated with ridge crests exacerbates the problems of stability.

An alternative to drilling for a first-order estimate of the vertical extent method for mapping sulfide deposits of SMS deposits may be provided by electrical methods. This method relies on detecting the small electrical currents generated within a metalliferous ore body. Some success has also been achieved by employing resistivity measurements from a submersible on the EPR at 13°N (Francis, 1987). Although insufficient measurements were taken to give an accurate estimate of the thickness of the ore body, the feasibility of this method was proven.

A method developed by the Canadians may possess some potential for delineating the vertical extent of deposits. Magnetometric off-shore electrical sounding (MOSES) records the magnetic field generated by a vertical current carrying long-line bipole extending from a ship to the seafloor. The characteristics of the generated field reflect the resistivity and the thickness of the strata in the seabed. This system has already been used to measure sedimentary thicknesses, and it was planned to deploy the system on the Explorer Ridge to attempt to determine the thickness of an SMS deposit.

It emerges from discussions that growing of the technology need for detailed proving of an SMS deposit are growing for better. Before commercial production is considered, detailed mapping would be necessary, probably involving a combination of electrical methods, grid-pattern drilling, and downhole geophysical methods. Technological developments likely to facilitate the sampling and proving of an SMS deposit include further incremental developments in the efficiencies of ROVs and deep-towed vehicles, particularly in the areas of power storage and telemetry (see also Chapter 9).

- Development of autonomous ROVs capable of operating efficiently for extensive periods on or near the seabed as discussed earlier
- Continuing refinement of deep-sea drilling systems capable of drilling into bare rock
- Development of efficient portable hard rock drills capable of operating at depths of 3000 m or more
- Refinement and development of electrical or other methods with the ability to produce an estimate of the thickness of a sulfide body without drilling
- Development of the technology to produce large-scale optical images of the seabed from 10–100 m above it.

If it is assumed that a large, high-grade SMS deposit has been located and proved by the methodology outlined in the previous sections, the next stage will entail mining and transport to the surface. Although commercial dredging on the continental shelf in depths of down to 50 m is now commonplace for material such as sand, gravel, and calcium carbonate, the technological problems encountered in mounting a similar operation in the deep-sea are formidable. At the present time, no commercial deep-sea mining operation exists, but this is not due to technological constraints, indeed hundreds of millions of pounds have been spent to prove the feasibility of recovering minerals from great depths.

The vast majority of this R&D effort has concentrated on manganese nodules, subspherical concretions varying from a few cms to around 20 cm in diameter, and containing significant amounts of manganese, copper, nickel, and cobalt. In the early 1960s, these nodules were shown to cover large areas of the deep seabed (average depth 5000 m), like pebbles on a beach, in total representing a resource far greater than known land deposits. Various industrial consortia were established to develop systems for their extraction, and these technologies still form the basic knowledge base regarding deep-sea mining.

The technological challenge of recovering minerals from the deep sea can be conveniently divided into three areas: a method of collecting the mineral from the seabed, probably incorporating primary screening; a means of transport to the surface; and a mothership or production platform (Table 10.3). Due to recent advances in dynamically positioned drilling ships and the plethora of exploration/production platforms developed for the offshore hydrocarbon industry, the latter factor can now really be regarded as a mature technology and there are unlikely to be any serious problems in designing a surface facility, tailor-made to the rest of the mining system. It is concerning the collector head and transport to the surface that the greatest technological challenges lie. The R&D activities of the late 1960s and early 1970s produced three basic solutions to these problems.

10.5.1 Continuous Line Bucket

This system is an attempt to adapt chain bucket dredging to the deep sea. It consists of a continuous wire or nylon rope supporting bucket dredgers at regular intervals. Figure 10.4 shows concept proposed for deep sea mining of manganese modules.

TABLE 10.3

Technological Requirements for Deep Sea Mineral Mining

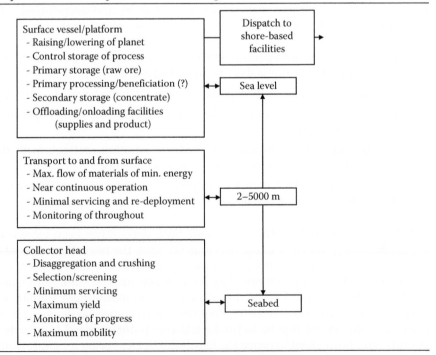

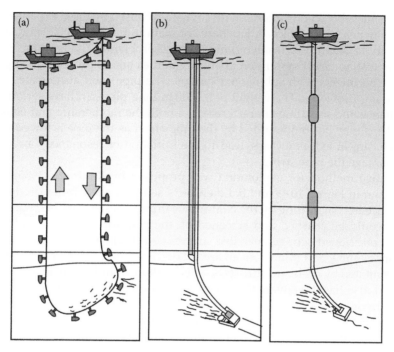

FIGURE 10.4
Continuous line bucket and other concepts for manganese nodule mining (a) continuous line bucket, (b) nodule sweeping trawler with hydraulic lift system, and (c) nodule trawler with intermediate station. (Drawn based on several available concepts.)

The wire and buckets form a continuous loop, dragged across the seabed by one or two ships. The buckets descend to the seabed, fill up with nodules, and return to the surface to empty their load. The technology is simple but energy-intensive. Trials have been conducted with this system, but success was limited, with problems occurring in snaring and tangling, and it is doubtful if this system will ever prove viable for a large-scale operation at depth.

10.5.2 Hydraulic/Airlift Methods

The majority of workable systems to date have employed some form of suction or airlift to transport nodules to the surface. Nodules on the seabed are collected by a suction head, often the size of a house. Collection may be by a passive head (shaped to scrape up the nodules) or an active one (employing mechanical means or high-pressure water jets to improve pick up efficiency). Most collector heads also employ some form of screening to minimize unwanted silt and mud in the slurry, and to exclude large blocks, which could cause pipe damage or blockage. The collector head can be either towed or self-propelled (as shown in Figure 10.4).

The resulting slurry of seawater and nodules is then pumped to the mothership through about 5000 m of pipe. This is usually achieved by submersible pumps at intervals along the pipe string. However, in the top third or so, some systems employ airlift. This consists of injecting compressed air into the pipe to lower the specific gravity of the slurry. The merits of airlift are debatable as the three-phase slurry (nodules, water, air) is very unpredictable and problems have been encountered with cavitation caused by rapid expansion of the air bubbles as they near the surface. A Japanese consortium has built a 200 m deep test tank at Tsukuba, in an attempt to resolve some of these problems. The tank has successfully been used to test an airlift system on artificial nodules and is presently being employed for similar experiments with submerged pumps. The pipe itself also creates many difficult engineering problems. The weight of a 5000 m long pipe can have serious effects on the performance and handling characteristics of both the mothership and collector head. These problems may be compounded by the drag effect as the pipe is towed through the water, which can, in extreme cases, lead to the initiation of resonance vibrations strong enough to damage the pipe string.

An alternative method for exploration was proposed by Kaufman (1985) and Zippin (1983) as shown in Figure 10.4a and b. Likewise, a self-propelled cutter dredge similar to what is practice in coal mining by the continuous mining system has been proposed.

Here ore would be crushed and screened on the platform and then transported to a dynamically positioned semisubmersible surface vessel, via an airlift pipe-string. When the collector assembly had mined out an area of ore, it would be lifted elsewhere, positioning being facilitated by on-board thrusters. This method would be capable of high production rates and also has the advantage of providing a stable work platform on the seabed, making the dredging operation more effective.

Self-Propelled Cutter Dredge: A variation on the above concept is to equip the platform with caterpillar tracks, making it easier to maneuvre; this would facilitate more efficient recovery of the deposit.

10.5.3 Chantier Sous-Marin (CSM)

The most futuristic solution conceived for nodule mining was that of the French national consortium, Association Francaise pour l Etude et la Recherche des Nodules (AFERNOD). This system was designed to replace the troublesome pipe string and mine the nodules with a fleet of autonomous ROVs. The original concept involved a fleet of free-swimming ROVs each weighing about 400 tonnes, and capable of lifting 100 tonnes of nodules per trip to the seabed. Lifting and submergence would be achieved by addition and discharge of ballast, probably the waste produced by primary processing aboard the surface facility. Collection of the nodules was to be achieved by two Archimedean Screws, which would also propel the unit along the seabed. Offloading of the nodules was to be by subsurface transfer, probably to some form of semisubmersible platform at the surface, which would also conduct primary processing of the ore.

Problems were encountered with power storage at the early stages of development of the CSM. Although new syntactic foam was developed to give the units buoyancy so much was needed to compensate for the weight of the batteries that the vehicles would have been far too large to be practicable. However, development of the system is continuing, being developed by the French consortium GEMONOD for ocean mining, in collaboration with Preussag AG, Germany. The original autonomous unit is still maintained, but no longer in the free-swinging mode. The current system would be a highly mobile self-powered collector head, collecting nodules and returning them to a static "conventional" pipe string at the seafloor.

All of the above concepts would, of course, require sophisticated backups, such as extremely accurate navigation and monitoring systems to ensure efficient nodule recovery rates. In addition, although CLB and hydraulic/airlift systems have recovered nodules from 5000 m, it is certain that many problems would be encountered in scaling up these systems to the size necessary for a commercial operation. The Red Sea Project is now at this stage and has already proved the viability of employing submerged pumps and a pipe string over 2000 m long to carry minerals to the surface. The Red Sea deposits are unconsolidated muds situated on a relatively flat topography, in contrast to the massive, hard SMS deposits in the very rugged terrain typical of spreading ridges. The technology employed in the Red Sea should be virtually directly applicable to other SMS deposits with one crucial exception, the collector head.

10.5.4 Other Options

The methods discussed in the previous section are all variations of the basic technology developed with a view to extracting manganese nodules. However, it may be more appropriate to design a new system geared specifically toward extraction of SMS deposits. Learning from experience and suggestions may be worthy of further investigation.

1. *Grab/Bucket Dredge*: It may be possible to develop a system utilizing large grab/ bucket dredgers akin to the one developed by Germany for sampling the deposit. The most likely configuration would involve a ship equipped with two dredgers, one in the bow the other in the stern. Each dredge could be equipped with TV cameras and possibly thrusters to increase mobility. Such a system would only be capable of mining surface deposits on a small scale but has the advantages of involving state-of-the-art technology, coupled with a high degree of mobility. Such a small-scale operation would be unlikely to be very profitable but may produce enough revenue to offset the cost of further developments, while at the same time greatly increasing current knowledge of the distribution and average grades of SMS deposits.

2. *The Shuttle Concept*: A system based on autonomous submersibles similar to those proposed by AFERNOD for mining nodules would have many advantages. A fleet of vehicles could be operated, ensuring continuous mining even when technical problems are encountered. The units would also be highly mobile and the accurate maintenance of position by the surface vessel would not be so crucial. However, as mentioned earlier, severe technical problems still have to be overcome before such a system can be considered viable. The major problems, at present, are the ability to store sufficient power within an autonomous vehicle for propulsion and crushing of the ore, and buoyancy problems. Control of the vessel would also present difficulties given present telemetry systems, and although this could be overcome by employing manned submersibles, the increased cost incurred by extra safety features would probably be prohibitive.An even more futuristic scenario would employ submersible mining units, which would feed underwater barges. When full these would be towed to shore by large manned submarines. This system would not only suffer from all the problems associated with operating shuttle miners but would introduce new technical difficulties associated with the towing operation. Though this method would obviate the need for a surface facility, it seems likely that the problems created in underwater power storage, buoyancy, and control would far outweigh the benefits of eliminating the surface vessel.

3. *Vent Capping*: It is in the nature of SMS formation that the vast majority of the metals contained in the exiting vent fluids are dissipated into the ocean, only a small fraction being incorporated into the sulfide chimneys/mounds. This fact, from the

earliest days, has fueled speculation into a method of tapping the fluids directly. Two basic methods have been proposed: capping of a vent and piping the fluids to the surface for precipitation, and capping by an autonomous collector left on the seabed.

The first of these proposals does not appear viable for several reasons. First, it would require the design and development of a totally new system capable of tapping several vents and conducting this material to the surface. There would be many difficulties associated with preventing the brines precipitating en route and blocking the pipe, and also with keeping the surface vessel accurately positioned for long enough to collect a reasonable amount of material. Such a system would also require constant delivery of high-grade ore-bearing solutions, and it is by no means proven that such vents exist.

The second method appears far more plausible, simply requiring some form of autonomous cap with an affinity for the desired metals to be lowered over active vents. The cap would then be released and recovered some time in the future with its load of metallic ore. This method would appear to present no great technical problems, though some device would have to be incorporated in order to prevent the cap becoming blocked leading to a dangerous buildup of pressure.

Cann (1980) estimates that the "average" vents emit fluids at 10–20 kg/s, which, at 3 ppm of copper, would only yield 1.8 tonnes/year. However, it is known that volcanic activity at ridge crests is episodic or even sporadic, and it must always be borne in mind that observations to date have only given us a snapshot of a limited number of locations at given points in time. What is now regarded as the "norm" may indeed be a gross underestimate (or overestimate). Until more is known about how these systems vary through time, it would be unwise to totally dismiss the idea of "milking" them as a steady, low-yield renewable resource. One way this may be achieved is by subsea automated systems, which are now commonplace in the offshore hydrocarbon industry. If a sufficient density of vents could be found at one locality it might be possible to tap many of these emissions and precipitate them at a central collection point from which they could be periodically recovered.

4. *Solution Mining*: Another method, which would obviate the need to transport large amounts of ore and waste material to the surface, is solution mining. This method involves pumping a lixiviant (an acidic or alkaline mixture of solvents) into a borehole so that it permeates the deposit, thus leaching out the desired metals. The metal-enriched mixture is then returned to the surface via another borehole for refining on board ship. This method has yet to be tried in a subaqueous environment, but several patents have been issued on systems that have enjoyed considerable success on land. One particular system developed by the Kennecott Corporation is claimed to have significant potential for the extraction of metals from SMS-type deposits. The method employs a two-phase ammonia leach solution containing tiny oxygen bubbles, which has extremely penetrative properties obviating the need for artificial fracturing of the ore body.

The technological problems associated with solution mining do not appear insurmountable. The major requirement would be accurate deep-sea drilling and a dynamically positioned surface vessel. The major disadvantage would concern possible problems in containment of the leaching solutions, which could render the process inefficient, and possibly lead to severe environmental pollution in the immediate vicinity of the operation. One recent development in this field is the development of systems for bioleaching (leaching by means of microorganisms), but to date the potential for employing this method on SMS deposits does not appear to have been investigated.

It is clear from the preceding sections that a great deal of research and development is necessary before a mining system for SMS deposits can become a reality. Possibly the most likely first-generation mining vessel would be one employing two or more scaled up versions of Germany (Preussag's) grab sampler. However, such a small-scale operation would be little more than a mechanism for gaining more information on what constitutes a "typical" SMS deposit and how widespread such deposits are. True commercial exploitation would have to work the deposit vertically to optimize yield. At present, the most likely scenario would appear to be a system utilizing existing hydraulic/airlift technology, combined with a purpose-built collector head, capable of mining the ore by cutting or ripping, followed by crushing to facilitate transport to the surface in the form of a slurry. Such a collector head would be a truly innovative design and would have to be capable of withstanding an incredibly hostile environment involving

- Near freezing temperatures coupled with the possibility of encountering extremely corrosive superheated fluids of several hundred degrees centigrade
- Pressures of 250 atm and more coupled with the possibility of seismic or volcanic events
- Extremely rugged topography, containing vertical scarps, fissures, volcanic edifices such as "lava pillars", etc

Not only would the system have to cope with all of the above problems, but it would also need to be highly maneuverable and extremely reliable and durable. The last factor would be crucial, as deployment and retrieval of such systems would inevitably be very time-consuming and consequently expensive.

An alternative to the hydraulic/airlift system could be the development of a radical new technological system designed specifically for SMS extraction, such as a shuttle system, vent capping, or solution mining. Innovative technological developments would be required in each case including high-capacity lightweight power storage systems for autonomous submersibles, improved methods of telemetry, and deep-sea navigation, enabling comprehensive monitoring and control of sophisticated autonomous submersible vehicles. The two following methods are necessary: (a) development of an efficient capping mechanism for active vents and the long-term monitoring of vents to assess volume and sustainability of emissions and (b) development of an environmentally safe, and efficient for use in solution mining.

All the above concepts would require a dedicated dynamically positioned surface vessel, or semisubmersible platform. This, however, can now be regarded as a mature field of technology and its development should present no real problems. It would be desirable for the surface facility to include some level of primary processing. Again, this is a mature technology requiring only incremental changes for use on SMS ores. Such an operational base would require large amounts of energy, and one possibility yet to be investigated could be the utilization of some of the abundant geothermal energy available at sites of SMS formation.

10.6 Mineral Processing and Refining

Processing appears to be the least problematic part of any proposed mining operation, as the ores found in SMS deposits are similar to the ores mined on land for many years in places such as Canada and Cyprus. Some problems were originally encountered with the processing of the Red Sea ores due to the unusually fine-grained nature of the desired minerals, but these have now been overcome. SMS typical of ridge crest deposits are far more similar to land-based ones. Work on the geotechnical properties of samples from the EPR at 21°N has provided valuable information necessary for designing suitable cutting/ ripping and crushing equipment, but it is important to bear in mind that all the samples were from superficial deposits and it is possible that subsurface deposits may differ in their geotechnical properties.

Primary concentration of the ores would most likely be achieved by some method of froth flotation. The subsequent stages of processing would vary according to the desired metals in a particular sample. Samples from the Juan de Fuca Ridge were subjected to chlorine–oxygen (Cl_2–O_2) leaching with excellent results by the USBM. The process succeeded in recovering 99% of the contained zinc and cadmium, 97% of silver, and 78% of copper. More recently, Carnahan suggests a standard treatment for MPS ores consisting of blending, crushing, and grinding, followed by bulk flotation to remove iron sulfide, selective flotation of copper and zinc sulfides, and final smelting to metal of the resulting concentrates. Finally, it may be possible to employ bacteria to release the metals contained in the sulfide ores. The ores are generally of a very porous nature, due largely to the action of worms and other fauna around the sites of SMS formation. Bioleaching has been applied in certain instances on a commercial scale, but tends to be a slow process. The speed of this method can be increased at higher temperatures, which gives rise to the interesting possibility that the thermopile bacteria present at vent sites may be genetically engineered to win metal from the ores at higher temperatures.

It would appear, therefore, that the technology now exists to process SMS ores efficiently without any radical new developments. However, ore processing is a site-specific activity and so a detailed proposal for beneficiation and processing can only be devised when details of the geotechnical properties, grades, and range of desired metals for a given deposit are available. As always, there is room for innovation, and bioleaching employing indigenous bacteria engineered to win desired metals from the ore is a possibility worthy of further investigation.

The Institute of Minerals and Materials Technology under the Council of Scientific and Industrial Research (CSIR) in India is conducting extraction techniques experiment from nodules, cobalt-rich crust, and seafloor sulfide. The MINTEC of South Africa conducts high-quality beneficiation and processing experiments.

10.7 Other Factors

The decision on whether or not to mine a particular marine resource is influenced by factors other than resource potential and extractability. Near-shore dredging of minerals has taken place for many years, but extraction in the high seas is a new phenomenon, and a consensus covering rights in international law has yet to be achieved.

In the absence of such agreement, it is highly unlikely that any deep-sea mining will take place. Any mining operation is also strongly influenced by existing alternative sources, even more so when a radical new technology is required for exploitation as in the case of SMS. Therefore, a thriving market in the relevant metal is essential for deep-sea extraction to become a reality. Finally, all modern mining operations must consider the environmental effects of projects. This is likely to be a crucial factor in any proposal to mine SMS, given the unique fauna normally associated with active vent emissions.

At present, two legal regimes exist governing a coastal state's rights regarding exploitation of seabed resources, one pertaining to the area within national jurisdiction, and the other covering activities in international waters.

10.7.1 Mining within National Jurisdiction

Two legal concepts can be invoked for the exercise of a coastal states right to exploit all minerals within 200 nautical miles of its coast, the exclusive economic zone (EEZ) and the continental shelf. The EEZ is a zone extending up to 200 miles from the baseline from which the breadth of the territorial sea is measured. In the EEZ, the coastal state has sovereign rights for the purpose of exploring and exploiting, conserving, and managing the mineral resources of the seabed and subsoil. The EEZ is a concept of recent origin and has been embodied into the 1982 United Nation Convention on the Law of the Sea (UNCLOS), which has not, as yet, entered into force. However, a large number of states have incorporated the EEZ into their national legislation and it can now be said to be part of international customary law.

The rights of a coastal state over minerals found on its continental shelf are governed by Article 1 of the 1958 Geneva Convention, which defines the continental shelf as "(a) the seabed and subsoil of the submarine areas adjacent to the coast but outside the area of the territorial sea to a depth of 200 meters, or, beyond that limit, to where the depth of the waters admits of the exploitation of the natural resources of the said areas; (b) to the seabed and subsoil of similar submarine areas adjacent to the coasts of islands."

This definition was somewhat vague in character and led to a series of disputes between coastal states. A more practical definition of a continental shelf was incorporated into the 1982 UNCLOS, which defines it as

> the seabed and subsoil of the submarine areas that extend beyond its territorial sea throughout the natural prolongation of its land territory to the outer edge of the continental margin, or to a distance of 200 nautical miles from the baselines from which the breadth of the territorial sea is measured where the outer edge of the continental margin does not extend up to that distance. (Article 76 para. 1)

Hence, the EEZ and continental shelf are synonymous in legal terms up to 200 miles, but the continental shelf, if it satisfies certain geological criteria, can extend jurisdiction beyond this up to a maximum of 350 nautical miles from baseline or 100 nautical miles from the 2500 m isobath. Another consequence of Article 76 is that areas of the deep-sea (off the continental shelf) now come under a coastal states jurisdiction if within 200 nautical miles from the territorial sea baseline. Seabed mining within national jurisdiction of a coastal state would be subject to the laws and regulations of that state but must not interfere with traditional freedom of the seas, such as the operation of ships and aircraft or the laying of submarine pipes and cables.

10.7.2 Mining in International Waters

According to UNCLOS, the seabed outside coastal state jurisdiction is termed "the area" and accounts for some 60% of the ocean floor. An international legal regime governing the exploration and exploitation in *the area* of all "solid, liquid and gaseous mineral resources at or beneath the seabed" is elucidated by UNCLOS, which incorporates the following general principles:

- The resources of *the area* are the common heritage of mankind and as such are not susceptible of unilateral national appropriation.
- Activity in *the area* shall be conducted so as to promote the peaceful and rational development of the resources and shall be carried out for the benefit of mankind as a whole.
- A new international organisation called the International Seabed Authority (ISA) participates in the exploitation of *the area* through its operating arm, the Enterprise, or through the granting of licenses for exploitation under its supervision by private contractors.
- The relevant technologies used by private operators must be transferred to the Enterprise on fair and equable terms and conditions.
- The minerals of *the area* must be developed in such a way as to promote a just and stable price and to protect developing land producers against adverse effects. There is a mechanism of control for the volume and value of the metals produced from nodule mining but not for other types of deposits such as SMS.

Following preliminary exploration, the mining contractor makes an application to the ISA for an exploitation area divided into two parts of equal commercial value. The authority selects one part to be reserved for its own mining activity. The second part is subjected to a licensing system whereby the contractor is given exclusive rights for the exploitation of a mine site by the authority, in return for complying with a series of rules and regulations, such as the following:

- The contractor is subject to production controls. He is obliged to transfer technology to the Enterprise (the operating arm of the ISA) on fair and reasonable commercial terms and conditions.
- The contractor is subjected to financial terms involving the payment to the authority of an administrative fee of $500,000 followed by an annual fee of $1 million.

Following the signing of the 1982 UNCLOS, the Preparatory Commission for the International Seabed Authority has been working on setting up the deep-sea mining regime and in particular on drafting a mining code.

Multilateral agreements such as the 1982 UNCLOS commonly take five to ten years to attract enough ratifications (in this case, 60) to enter into force. To avoid such delay and in view of the dissatisfaction of a number of Western nations (in particular the United States, UK, and Germany) with the mining provisions contained in the 1982 UNCLOS, a group of "like-minded states" have established a Reciprocating states Agreement (RSA or "mini-treaty") to permit and regulate mining before the 1982 UNCLOS enters into force. This regime results from the formulation and adoption of similar national laws governing activities in *the area*, which are coordinated by a series of international agreements aimed at reducing disputes between the participating countries.

Russia also has its own national legislation, but it is not a party to the Reciprocating States Agreement. This agreement has been presented by the participating states as an interim agreement pending implementation of the 1982 UNCLOS. Although it is designed to assure mutual recognition of participating countries' claims, it does not prevent states outside the agreement from claiming overlapping areas. The basic legality of this agreement is also challenged by the "Group of 77," a collection of over 100 nations who claim that the concept of deep-sea minerals as being the common heritage of mankind is already part of customary law, and as such unilateral claims are illegal. It is therefore likely that any attempt to mine *the area* under national legislation would provoke stiff international opposition. Another problem with these national legislations is that they were drawn up with the exploitation of manganese nodules as their objective and so may not be directly applicable to other deep-sea deposits such as SMS.

In summary, the legality of mining SMS deposits must be viewed in two contexts. First, mining of deposits within 200 nautical miles of a coastal states' territorial sea baseline or within the continental shelf as defined in UNCLOS, could legally take place, subject only to the terms and conditions laid down by the relevant government. In contrast, mining of any minerals in *the area* would be subject to significant legal uncertainties and would be open to challenge by third party states on the grounds that it violated international customary law. Mining of SMS in *the area* under existing national legislations could also be challenged on the grounds that these statutes were formulated specifically for manganese nodules and are not applicable to any other resource.

References

Cann, J. R., 1980. Availability of sulphide ores in the ocean crust. *J. Geol. Soc. London*, 137: 381–4.

Francis, T. J. G., 1985. Resitivity measurement on an ocean sulfide deposits from submersible Cyana. *Mar. Geophys.*, 7: 419–437.

Francis, T. J. G., 1987. Electrical methods in the exploration of sea floor mineral deposits. In: P. G. Telki. *Marine Minerals Edt.* Riedl, London, pp. 413–419.

Hammett, D. S., 1985. Ocean Drilling Program Vessel/equipment capabilities. *17th Annual Conf. Offshore Technology*, Houston, 287–294.

Hekinian, R. and Fouquet, Y., 1985. Volcanism and metallogenesis of axial and off-axial structures on the EPR near 13°N. *Econ. Geol.*, 80(2): 221–249.

Kaufman, R., 1985. Conceptual approaches for mining marine polymetallic sulphide deposits. *Mar. Technol. Soc. J.*, 19(4): 50–56.

Kunzendorf, H., 1986. *Marine Mineral Exploration.* Elsevier Sci. Publ. Amsterdam, p. 300.

Rona, P. A., 1984. Hydrothermal mineralization at seafloor spreading centre. *Earth Sci. Rev.*, 20: 1–104.

Rona, P. A. and Scott, S. D., 1993. Preface 1933–1976. In: Special issue on seafloor hydrothermal mineralization: New perspectives. *Econ. Geol.*, 88(8): 1993–2295.

Roonwal, G. S., 1986. *The Indian Ocean: Exploitable Mineral and Petroleum Resources.* Springer-Verlag, Heidelberg, Vol. XVI, p. 198.

Sillitoe, R. H., 1982. Extensional habits of rhyolite hosted marine sulfide deposits. *J Geol.*, London, 10: 403–407.

Zippin, J. P., 1983. Draft Environmental Impact statement, Proposed Polymetallic Sulphide Minerals Leasing Offer, Minerals Management Service, US Department of the Interior, Dec. 1983, 580p.

Useful Websites

2H Offshore – www.2hoffshore.com.
GNS Science – www.gns.cri.nz.
JSMLS – www.jsmls.com.sg.
Koppel Offshore and Marine Technology Centre.
Krypton Ocean Group – www.kriptonocean.com.
Marine Space. www.marinespace.co.uk.
National Institute of Ocean Technology website.
Nukurangi. www.niwa.co.nz.
Ray Wood. www.rockphosphate.co.nz.
Royal IHC Mining. www.ihcmarine.com.
RPS Group. www.rpsgroup.com.au.
World Ocean Council – www.oceancouncil.org.

11

Ocean Environment and Pollution

11.1 Environmental Issues

In vast size of the world ocean, the deep sea beyond the continental shelf, covering more than 300, 10^6 km^2, is a huge water mass. But so far humanity has used this vast reservoir as a dumping area with no regard for the environment. In the shallow and coastal area, signs of high degree of pollution can be seen clearly, especially around megacities such as Mumbai (Glasby and Roonwal, 1997). In the deep sea, all experimental operations using trawls, dredges, covers, and traps during the past 150 years have been going on. As the world looks to the oceans for resources and mining both for hydrocarbons and hard minerals, as also for intense fishing, it is time to examine the impact of human intervention on marine environment. One can see signs of dissolution or bleaching of coral reefs on the acidification of ocean water. This also affects the biodiversity of the ocean and the earth. It is important to study both the mega-scale and minor-scale impact on marine environment due to mining- and drilling-related issues (Amos et al., 1973).

In this chapter, the environmental issues are therefore enumerated on the basis of likely mining of hard minerals in the oceans, including the Indian Ocean. Scientists need to decide the right scale of monitoring of such ocean observation experiments. For the manganese nodule, basic information on environmental effects are generally known (Thiel et al., 1992) and in the Indian Ocean as well (Sharma et al., 2001).

Environmental changes in the seashore zone can occur most dramatically as a result of eustatic sea-level changes. Therefore, what was previously a shallow marine environment becomes dry or vice-versa. This is shown by the link between "S" and "E" in Figure 11.1 It can similarly be linked to tectonism through relative sea-level changes. Tectonics can also affect the depositional environment in other ways. Uplift in the coastal hinterland can result in a marked increase of sedimentation, producing sea-level progradation of the coastal zone, siltation of coastal lagoons, or transformation of the shelf from carbonate-dominated to clastic-dominated. This type of relationship is shown in Figure 11.1 by the link between "T" and "E."

Not only does tectonics play a direct role in the formation of marine nonliving resources, but it can also indirectly play a role through its influence on sea-level changes, marine environments, and climate as shown in Figure 11.1. Tectonic process can be very important for the formation by marine nonliving resources.

They can take place on scale of a few years to many million of years and can affect a relatively small part of the earth's crust in the case of localized uplift, for example, on vast parts of the globe in the case of plate movements.

Eustatic sea-level changes and environment are interlinked changes in the sea level that occur on a variety of scales ranging from a diurnal tidal change to a very major sea-level

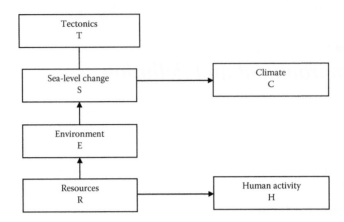

FIGURE 11.1
The set mechanism and some of the potential "feedback."

change associated with the development of ice caps, and uplift or sinking of major parts of the earth's crust. The major changes in sea level that are global in extent are referred to an eustatic. In recent years, earth science has greatly improved on documentation of the earth's eustatic sea-level history, largely through the efforts of the petroleum industry, particularly through interpretation of seismic profiles. Thus, ancient shoreline features such as coral reefs and beach ridges have been delineated.

Environmental changes can also be linked to climate changes. Climatic changes can affect rainfall, which may affect freshwater input into the coastal zone, in the form of surface water and groundwater. This can be simulated to some extent by damming river, and the resultant effect of this on the coastal zone can be very considerable. This is an example of the H to E links shown in Figure 11.1. Local and regional fall in water temperature can bring about a decrease of coral growth. An increase in the storms and related features on the energy regime can have a very marked effect on the distribution and concentration of near-shore and continental shelf sands and on the formation or destruction of beach ridges. Above all, anthropogenic or human-made change are important. Society has the wisdom and capacities to manipulate these changes keeping in mind the long-term implications and rules of nature. This will pave be the way for sustainable development for the welfare of humans.

Environment effects of deep-sea mining is an important aspect that requires immediate attention and assessment because the deep-sea minerals have attracted a great deal of attention as a promising resource in the future, as also many industrialized nations have conducted exploration activities with great enthusiasm, besides the developing nations, in the Indian Ocean region.

The research and development of manganese nodule mining and mining tests were completed in late 1970s as mentioned before. On the other hand, due to increasing concern from all over the world about the preservation of the environment and depreciation of base metals such as copper, lead, and zinc, as well as rare metals such as nickel, cobalt, and others, the development of manganese nodule research has been seriously conducted by many nations, more so by the nations with a growing economy. At the same time, since industrialized nations were requested to solve the problem of the destruction of the ocean environment and thus the deep-sea mineral resources, they have been conducting marine environment surveys, collecting valuable data and information on the marine environment such as on the physical, chemical, and biological condition and the marine ecosystem. Table 11.1 gives an outline of different factors that need attention for such an evaluation.

To understand the marine environment, data and information concerning the surface layer, underwater, and the seabed must be collected. At the surface layer of the ocean, physical phenomena such as waves, ocean currents, and others impact a number of areas. A great number of living things such as fish, plankton, and others exist underwater, particularly near the surface layer. Since their living conditions depend on the concentration of oxygen, nutrient salts, and other elements are dissolved in the sea water, it is necessary to measure the dissolved and suspended substances to understand the chemical phenomena. It is also necessary to study the ecology of zooplankton, phytoplankton, fish, and large animals such as whales. On the seabed, it is necessary to conduct research in order to truly understand the ecology of benthos including bacteria and protozoan and others.

To understand the relationship between resedimetation and biological reaction, artificial impact experiments are being conducted by the United States, Germany, and India. The former, named Benthic Impact Experiment (BIE), began in 1991 and is being performed by the National Oceanic and Atmospheric Administration (NOAA) of the United States in the north equatorial Pacific Ocean. The biter named disturbance and recolonization experimentation in the deep south Pacific Ocean (DISIOL) was started in 1989 and is being conducted in the south equatorial Pacific Ocean by a scientific group from Hamburg University, Germany. The Metal Mining Agency of Japan (MMAJ) carried out environmental studies. These studies, commenced in 1989, were conducted in association with NOAA and include an experiment named the Japan Deep Sea Impact Experiment (JET). The Central Indian Ocean manganese nodule mining environment experiment was conducted by India as an INDEX experiment (Sharma, 2001, 2005).

India has now conducted her program of environmental impact assessment (EIA) in the Central Indian Basin (CIB), the area put for claim by India to the International Seabed Authority (ISA). An evaluation of these experiments is awaited. These experiments are mainly on the pattern of JET, but preliminary results have been published (Sharma, 2001, 2005).

There was an epoch-making research program on the marine environment, which was known as deep-ocean mining experiment study (DOMES), which began in 1975, to

TABLE 11.1

Marine Mining Impact Evaluation

Mining impact evaluation	Biological response	Macrobenthos
		Mesobenthos (Density and sedimentary bacteria)
	Impact mechanism	Organic carbon + nitrogen
		Sedimentary bacteria
	Impact Area	Sedimentary matter quantity
		Current velocity and direction
	Recovery process of benthic community	Macrobenthos
		Meiobenthos
		Sedimentary bacteria
	Impact level	Organic carbon and nitrogen
		Calcium carbonate concentration
		Opal concentration
		X-ray photography
	Background factor	Water content, grain size
		Nutrients, pore water, interface water

evaluate potential of sea environmental effects from deep ocean mining. The DOMES program focused on studies relating to short-term (one week), near-field (5–10 km from the point of operation) physical, chemical, and biological effects of deep-sea mining. DOMES consisted of two phases. DOMES-I characterized the base baseline condition within the 13 million square kilometers of the ocean area where the industry had indicated that initial mining would begin. DOMES-II monitored the effects from industrial pilot-scale tests within the same area.

11.2 Environment and Tectonics and Eustatic Sea-Level Changes

Let us briefly examine each of the three factors that influences the deposition of mineral resources in the nearshore and continental shelf. However, it must be recognized that these are not totally independent factors. There are feedback mechanisms that result in the development of various interrelationships and in some cases the formation of nonliving resources.

Tectonics is described in the American Geological Institute (AGI) dictionary as the "study of the broader structural features of the earth and their causes." Processes such of volcanic activity, earthquake, faulting, folding, uplift, and down wrapping all fall within this preview of tectonics. All these processes could be regarded as a manifestation of plate tectonics and the attendant mechanisms of plate separation, collusion, and subduction. Tectonic processes are of fundamental importance to the formation of oil and gas fields and mineral deposits both onshore and offshore. Mineralization, commonly associated with "black smoker," occurs on midoceanic ridges where the ocean floor is being generated. Deformation the continental margin can produce deep structures for oil and gas. An increase in the geothermal gradient under the continental margin can result in the sediment entering the "oil window" and generating hydrocarbons. Onshore, the formation of a mountain range can result in the exposure and erosion of rocks containing gold, ilmenite, or diamonds, which are then transported down the rivers into the coastal zone to form marine placer deposits.

Not only does tectonics play a direct role in the formation of marine nonliving resources, but it can also indirectly play a role through its influence on sea-level changes, marine environments, and climate as shown in Figure 11.1.

Tectonic process can be very important for the formation by marine nonliving resources.

They can take place on scale of a few years to many million of years and can affect a relatively small part of the earth's crust in the case of localized uplift, for example, on vast parts of the globe in the case of plate movements.

While all nonliving shallow marine resources are influenced to varying degrees by eustatic sea-level changes, they react at different temporal scales. For coastal zones as resource (CZAR), it is necessary to understand record of sea-level change over 10^0 to 10^4 years. For marine placers, we need to be concerned mainly with sea level changes over 10^3 to 10^6 years. For phosphorites, it is necessary to understand the sea-level record for 10^6 to 10^8 years.

For many nonliving resources such as placers, the effect of a relative rise of sea level and a eustatic rise of sea level are essentially the same in terms of influence on the distribution of resources. However, in the same way, notably phosphorites, there appears to be wide scale, perhaps even global episodes in phosphogenesis that in some cases can be related to eustatic sea-level rises. Therefore, in such cases, in order to use sea-level changes predictively for locating deposits, it is necessary to separate out relative and eustatic sea-level changes.

11.3 Environmental Effects Due to Mining

Deep-sea mineral resources have attracted a great deal of attention as a promising resource in the future and many developed countries have enthusiastically conducted exploration activities. The research and development of manganese nodule mining and mining tests at sea were completed in the late 1970s as mentioned before. On the other hand, due to increasing concerns from all over the world about the preservation of the environment and depreciation of base metals such as copper, lead, and zinc, as well as rare metals such as nickel, cobalt, and others, the development of manganese nodule research was forcibly postponed. At the same time, since developed countries were requested to solve the problems of the destruction of the ocean environment and thus the deep-sea mineral resources, they have been conducting marine environment surveys, collecting valuable data and information on the marine environment, including physical, chemical, and biological conditions and the ecosystem.

In order to understand the marine environment, data and information concerning the surface layer, underwater, and the seabed must be collected. At the surface layer of the ocean, physical phenomena such as waves, ocean currents, and others affect wide areas. A great number of living things such as fish, plankton, and others exist underwater, particularly near the surface layer. Since their living conditions depend on the concentration of oxygen, nutrient salts, and other elements dissolved in the sea water, it is necessary to measure the dissolved and suspended substances to understand the chemical phenomena. It is also necessary to study the ecology of zooplankton, phytoplankton, fish, and large animals such as whales. On the seabed, it is necessary to conduct research in order to truly understand the ecology of benthos, including bacteria, protozoa, and others.

JET included two cruises in the summer of 1994, both of which were conducted in association with the Central Marine Geological and Geophysical Expedition of the Russian Federation (CGGE) aboard CGGE's vessel the RN *Yuzhmorgeologiya*. JET was designed to evaluate the impact of resedimentation on the deep-sea benthic community and consisted of three phases. Phase 1 was a baseline study to confirm the predisturbed condition of the benthic community (density, fauna, composition, environmental factors). Phase 2 was intended to create an artificial resedimentation condition using a mechanical benthic disturber. Phase 3 was aimed to observe and analyze the biological reaction associated with the resedimentation and to understand the recovery process. Based on these phases, the biological impact of sedimentation will be assessed both spatially and temporarily.

There was an epoch-making research program on the marine environment, which was known as Deep Ocean Mining Environmental Study (DOMES). It began in 1975 to evaluate the potential at-sea environmental effects from deep ocean mining. The DOMES program focused on studies relating to short-term (one week), near-field (5–10 km from the point of operation), physical, chemical, and biological effects of deep ocean mining. DOMES consisted of two phases: DOMES I characterized the baseline conditions within the 13 million square kilometers of the ocean area where industry has indicated initial mining would occur; DOMES II monitored the effects from industrial pilot-scale mining tests within the same area. The program also broadly characterized the region of potential mining, since so little is known of the deep-sea ecosystem where mining is expected to occur. This was indeed a leading and epochal program and provided much important data and information to the world (Bischoff and Piper, 1979).

11.3.1 Potential Effect of Nodule Mining

The currently proposed method of collecting nodules from the seafloor shall certainly be accompanied by certain levels of environmental impact. The impact of its severity shall differ from one mining site to another mining site, and the mode of mining method adopted. At present, a complete level of the likely impact is not fully known because of inadequate data.

A variety of expected environmental effects due nodule mining has been identified in the Indian Ocean among others by Thiel (1991), Sharma (2001, 2005). A few major results of suggest that no significant damaging effects of long-term, wide-ranging consequences. In the Indian Ocean, it has been demonstrated quite well (Sharma, 2001, 2005) that certain potential problem areas remain, such as with respect to sediment plume generating directly from the mining activity on the site and from the discharge of water containing sediment and nodule fluxes (Amos and Roels, 1977).

As shown in Figure 11.2 direct impacts within the mining collector route shall happen and has to be in a way accepted if and when mining is carried out. The abyssal benthic fauna in the mining area shall be distributed and lost (Thiel et al., 1993). How long it will take to recolonize the distributed need to be investigated.

In Figure 11.3 is shown another aspects that needs to be known is the extent and accuracy with which a nodule collector is capable of picking nodules with minimum disturbance. This is an engineering challenge.

Most of us who have done the collection of nodules with devices like dredge grade sampler, box cover, and free fall grab have seen sediment sticking to nodules, still remaining on nodules, in spite of almost 4500 m of rinsing in the seawater column. Some studies show that despite the damage related to likely mining activity, no species loss is anticipated (Sharma, 2011, Thiel, 1992).

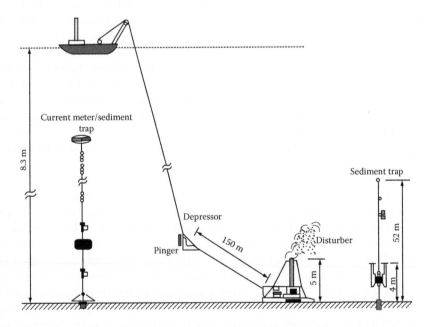

FIGURE 11.2
Schematic arrangement for disturber towing. (Based on Sakasegawa, T., and Matsumoto, K., 1997. Deep sea manganese nodule mining. Marine Industrial Technology, UNIDO, 1–18.)

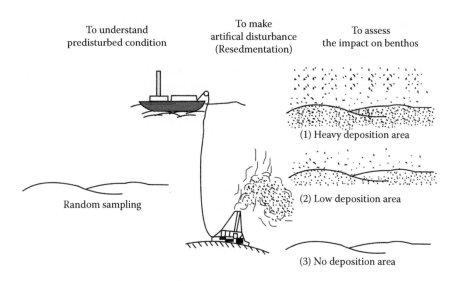

FIGURE 11.3
Sediment sampling was conducted to evaluate the effect of resedimentation on the benthic community by comparing predisturbance samples with those collected after disturber towing. (Based on Sakasegawa, T., and Matsumoto, K., 1997. Deep sea manganese nodule mining. Marine Industrial Technology, UNIDO, 1–18.)

Let us understand that in the deep-sea nodule mining, the sites seem to fall into the range of 20 to several hundred km² and are often likely to be separated from earth after based on nodule abundance, ore grade, and topography and other regions. This is nature's control on impact on marine life due to likely mining.

Although such reasoning may suggest the environmental consequence being confined to a relatively low level, and to vary from site to site, yet this aspect needs further thorough research.

In order to understand the likely huge transport of plume and associated transport of nodule material during the mining process, we need to fully understand the degree and level of impact. This shall help in recolonizing and sort of eco-restoration of the mine sites. As a case, we know that a typical mine site may show a nodule coverage of 20% and if all the associated sediment is rinsing off, before the nodules are sucked into the riser pipe string, it would leave a huge quantity of sediment being mobilized. This is only an indicator; the details are available in different publications on the theme.

11.3.2 Potential Effects of Cobalt-Rich Crust Mining

The single most influence of cobalt-rich crust mining is within the mine site. As compared to nodule mining, in this case, the sediment disturbance shall be of a much lesser level and influence. This is because unlike nodules, the cobalt-rich crusts are attached to the substrate material and on top of seamounts and then in a far-less sedimented site. Consequently, mining operation shall lead to sediment disturbance of no real consequence. However, here during the mining process or activity, the substrate is disturbed and during onsite separation by any system such as solution mining. The rejected substrate material would be dumped on-site and may eventually get distributed and redeposited in the area. Here, the disturbance to fauna at the mine site is to be expected. It therefore has to be accepted as a part of the mining activity.

Whichever method of mining is adopted, eventual transport of ore through the water column approximately ranging between 700 and 3000 m needs to be investigated for better planning. Here we need to accept that solution mining would not produce a large-scale devastating discharge anticipated as in case of nodules through a closed pipe string, and as enumerated in nodule-mining options, there would only a little disturbance to the environment and should be of a very low consequence.

However, similar to nodules, the mining of cobalt-rich crust has to be carried out only when it is likely to be economically viable. Here, again it is the topography that plays a greater role than in case of nodule mining.

11.3.3 Potential Environmental Effect of Metalliferous Muds of Sediments: A Case Study of Red Sea

In the Red Sea, the total environmental system is controlled by special condition such as a limited inflow of both the freshwater and a very restricted exchange with Gulf of Aden, Arabian Sea, and the main Indian Ocean. Further, due to arid condition in the region, hot winds always blow. And because of strong sun, an intense solar irradiation is experienced by this region. Consequentially, the ecology is of a special nature here. In the coastal areas, there are coral reefs. Due to this special condition, the ecology and biodiversity is of a specific nature, adapting to the special situation. High temperature leads to rapid evaporation, thus increasing salinity. Furthermore, as we have seen, the metal-rich sediment and mud occur in different "deeps" in the rift zone. This gives a unique feature to the Red Sea deposits.

The Red Sea Commission conducted some useful environmental impact studies during the course of exploration campaigns in the Red Sea (Nawab, 1984, Thiel 1991).

The environmental studies incorporated the physical, biological, and geophysical parameters for this special situation. It examined the possible effects on the ecology, if these metal-rich mud is mined. Details of the results can be seen in Thiel (1991).

11.3.4 Potential Environment Impact Related to Seafloor Massive Sulfide Mining

The consolidated SMS deposits should represent a significant source of Cu, Zn, Pb, Ag, and Au. The important land-based deposits are mined world ores—the large deposits in Canada, Japan, Oman, Cyprus, Peru, Chile, Australia, and other places. They are often designated as ancient analogs of the modern seafloor SMS deposits (Amann, 1985; Degens and Ross, 1969; Rona, 1983). The details have been described earlier.

A few seafloor SMS deposits have been considered to be potentially of sufficient quality for commercial mining. The mining method would center around the likely possibility of (a) scraping of ore from a surficial deposits, (b) excavating ore from the substrate in an open pit, (c) fluidizing the ore in a solution or slurry, (d) and tunneling into ore and removing from beneath the surface (Cruickshank, 1990). These methods may be operated in a combined way.

The type of SMS and their special nature of occurrence present some broad evaluation for potential mining. Similar to land-deposits of sulfides, sea seafloor SMS are also three-dimensional bodies. Environmental effects resulting from mining of SMS shall therefore be of limited scale, especially as compared with manganese nodule, which are two-dimensional deposits.

It looks like seafloor SMS deposit mining shall thus be similar to terrestrial sulfide deposit mining that we are familiar with. As the future SMS mining moves from concept to reality, there would be more research needed to evaluate the potential mining effect Learnings from land mining experience and adopting ways to minimize the likely effect of mining on seafloor SMS are likely to happen.

It is clear from preceding sections that at present, it is uncertain what method, if any, would be employed to mine SMS deposits. Consequently, it is hazardous to attempt to quantify the likely environmental effects of such an operation. Extensive work has been conducted on the possible environmental effects arising from a manganese-nodule mining operation, but such work is of limited value regarding SMS deposits, due to the fundamental differences in the biota associated with abyssal plains and vent sites. Two major environmental studies are pertinent to SMS exploitation: the Red Sea project and the Demonstration Environment Impact Studies (DEIS) prepared for the proposed Gorda Ridge leasing.

From its inception, the Red Sea Commission determined that the exploitation of the metalliferous muds should not have any deleterious effects on the environment. Throughout the Atlantis II project, great emphasis has been placed on monitoring the environment for adverse effects, and the commission pledged to cease operations should any such effects be proven. It was foreseen, at an early stage, that problems could arise regarding the discharge of fine particulate matter from primary processing of the ores onboard the ship. This could affect the environment by blocking out sunlight, contaminating the water column with toxic metals, or interfering with the feeding of siphonophores such as corals. In an effort to ameliorate these effects, the original discharge pipe vented its products at a 400 m depth, a process known as "downshunting."

It subsequently became apparent that this depth was too shallow and that the cloud of particles was being dissipated over a large, biologically active area. Eventually the outlet will be extended to 1000 m depth in such a position that the particulates should settle on the rift axis outside the deep, where very little fauna exists. An additional danger is presented by the discharge of water-soluble toxic substances along with the tailings; however, the results so far from laboratory experiments have been contradictory. Although environmental studies in the Red Sea have been wide-ranging, the assessment of long-term effects remains a fundamental problem. Indeed environmental studies themselves need to be long term to provide valid results on possible chronic effects. In the case of the Red Sea, it may take five years of monitoring a commercial operation before any adverse environmental effects are displayed, and by that time, it is likely that considerable damage to the environment would have already occurred.

The problems of assessing the environmental risks associated with any mining operations on the Gorda Ridge are even greater. The fauna associated with active vents is so new to science that it is virtually impossible to predict the effects of a particular mining scenario. The Gorda Ridge Demonstration Environment Impact Studies (DEIS) considers the likely environmental impacts in three main areas: the physical environment, biological resources, and socioeconomics. Regarding the physical environment, the report suggests that water quality could be affected by sediment plumes (from the returned seawater), heavy metal pollution, and accidental spillage of fuel and/or ore. The report concludes that environmental degradation would be localized, short term, and have a very low impact. The same conclusion was reached about any impacts on air quality, mainly arising from onshore concentrating facilities. Finally, there is a slight risk of low-level radioactive contamination. This is due to the fact that the area contains two low-level radioactive waste dumps, and it is possible that the containers could have ruptured or drifted, although physical contact with them would be extremely unlikely.

The west coast of the United States of America has a rich and diverse flora and fauna, and the DEIS concentrates on eight aspects of this biota: primary productivity, marine mammals, marine and coastal birds, estuaries and wetlands, areas of special concern and marine sanctuaries, endangered species, fisheries, and benthos. Of these the effects on primary productivity are expected to be small, which is a reasonable assumption considering the effects will be dispersed over a huge area in relation to the localized scale of any mining operations.

The effects on marine mammals and coastal birds due to collisions and construction of onshore facilities are described as low to moderate. Impacts on wetlands could be moderate to high since accidental release of fuel and/or ore is likely. The same also applies to areas of special concern and marine sanctuaries, which might either be unaffected or could be devastated. The effects on endangered and threatened species in the area are described as quite low for species that are not common in the lease area; only a moderate number of species visit the lease area.

It is important to continue to work on the fisheries and benthos, because these could well prove crucial in deciding whether environmental consequences will inhibit SMS mining in the area.

Fishing is a very important industry along this stretch of the coast, particularly in the small towns and villages. The DEIS highlights three causes of possible impact: suspended particulate matter (sediment plumes), release of toxic metals, and explosives (for seismic surveys or mining).

Sediment plumes generated during mining operations are likely to have an adverse effect on the ecosystem. In particular, the photosynthetic phytoplankton community would suffer as increased turbidity decreased light penetration. In turn, this would adversely affect all other components of the ecosystem, which all are dependent on phytoplankton. Effects could be minimized by employing downshunting, as in the Red Sea.

Toxic metals will also affect phytoplankton; however, the zooplankton community will suffer the most. The zooplankton community consists of microscopic consumers (i.e., they are nonphotosynthetic), some of which exist in the zooplankton throughout their life cycle, while others are the larval forms of fish. For both permanent and temporary members of the zooplankton, copper and zinc concentrations of 102–103 ppb (parts per billion) are lethal. Lower concentrations can induce morphological and developmental defects and reduce the population size by impairing reproduction. Although adult fish can withstand much higher concentrations (23,000 ppb zinc for steelheads), over a period of time, heavy metals may accumulate in their tissues due to the contamination of their prey. The main commercial surface feeding fish in the area (salmon and albacore tuna) are migratory, so "bioaccumulation" should not be too great a problem, as they are merely passing through. Also as the total area affected by metal pollution should be less than 1%, it is hoped toxic damage will be minimized.

Finally, while the report recognized that underwater explosions have a severe effect on all fauna in the immediate vicinity, it argued that since these effects will only be intermittent and restricted to a radius of a few hundred meters, at most, the effect on fisheries should be minimal. The report advised avoiding the use of seismic explosives when marine mammals are in the vicinity, as they can be injured due to their sensitive hearing.

It is crucial for prospective operators to minimize deleterious effects on the local fisheries. Since fishing is labor-intensive, any observed effects would elicit strong environmental and political opposition to the mining operations. Although any environmental impact would probably be slight, it could jeopardize any future projects, due to the strength of local public opinion.

The final biological resource considered by the DEIS is the benthos or bottom-living communities. Of these, the communities most at risk are the unique life forms associated with the hydrothermal vents. Large tube worms in the vent are common, and giant clams, crabs, fish, and anemones proliferate the vast majority of these being new to science. Such communities are associated with all the active hydrothermal vent discoveries to date. Indeed, so consistent is the association between these communities and hydrothermal vents that the presence of beds of clams has proved a useful indicator of hydrothermal activity. As mentioned previously, these communities are not light-dependent but rely on chemosynthetic bacteria able to synthesize organic matter from the sulfides emanating from the vents. They would be affecting the communities due to SMS mining.

Apart from direct effects of the mining operation, indirect effects of the sediment plume produced are likely to damage the ecosystem. This plume will extend approximately 100 km downstream from the mining operation, covering a few thousand square kilometers. As sedimentation is normally very low, many organisms, particularly filter feeders, will be adversely affected by this sediment over a large area. Corals, which normally live in areas of low sedimentation, are killed by small increases in sediment, which blocks their feeding apparatus and the same will apply to many members of the vent communities. Effects from toxic metals should be less pronounced on these organisms than most, as they have evolved in an environment naturally rich in these substances.

The environmental impact may be less drastic than it appears. Since vents are ephemeral phenomena (ordering tens of years), it seems likely that vent communities will have the ability to rapidly colonize new habitats as they become available. The exact mechanism of colonization is, as yet, unknown. It is imperative that research is concentrated on these communities in order to minimize damage. Again it is vital that any proposed mining operation should make every effort to minimize effects on these communities, for example by concentrating operations on inactive vents.

The main deleterious socioeconomic effects are likely to be on commercial fishing. Damage could also be felt by local subsistence economies dependent on salmon fishing. If sufficient safeguards are introduced, coupled with constant pollution monitoring, effects should be minimal. On a larger scale, the overall economy of the area should not be significantly affected by the exploration and development phases. However, if full-scale mining operations begin, some benefits should become apparent to the nearest coastal communities. There will be an increase in revenue to the area, and some jobs will be created, though not many, as mining will not be labor-intensive. These effects would be most apparent if onshore facilities were established in small communities. The actual scale of any socioeconomic impacts can only be assessed once specific operations are considered, and the decisions taken regarding location of onshore supply and processing facilities. Effects on recreation and tourism should prove minimal, as operations would be located far enough from shore to have no harmful effects on the aesthetic appeal of a stretch of coastline.

Geographically, the Gorda Ridge would appear to be a fairly typical site. Similarly, although the biology of the area is particularly rich, this is not unique to this particular deposit. All the likely SMS mining sites to date have occurred in areas that contain an equally diverse faunal assemblage. The vent communities appear to be ubiquitous at SMS sites; thus the conclusions of the DEIS would again seem appropriate for other locations. Socioeconomic factors could vary greatly for different SMS mining operations. Future projects are possible off the coasts in a number of countries including the United States, those in South America, and Japan. Each of these areas possesses widely differing political ideologies and cultures, and so the socioeconomic effects require individual assessment.

11.4 Potential Environment Impact Related to Phosphorite Mining

Offshore phosphorite nodules were first dredged on the continental shelf of South America during the *Challenger* expedition (Murray and Rennard, 1891). Since then, near-coastal phosphorite deposits have been located on the shallow seafloor and on the top of the seamount in a review (Cook, 1988). The southwest continental shelf of India has shown phosphatic muds (Roonwal, 1986) in a review as well as the southwest of African coast (Birch, 1980).

Economic potential in marine phosphorite is from Agulhas bank (140 million tons of P_2O_5 Baturin, 1982) and Namibian Coast (Bremner, 1980). The most attractive marine phosphorite is located in the Chatham rise of New Zealand. So far, no marine phosphorite has been mined. Therefore, no specific mining methods have been developed and related environmental impact studied. However, it is likely that air lift system of cutting wheel method, or continuous line bucket could be adopted in mining. The potential mining could be even a hybrid collector system if future so demands.

Like other mining activity, phosphorite mining could also impact marine environment and create a near-bottom plume quite like we have talked about in manganese nodule mining. Therefore, environmental aspects need an investigation to recommend minimizing the adverse effect.

11.5 Wastes in the Deep Sea

The oceans have received waste material from population, which settled along the coast line. It has now becoming an important issue. As we notice that during the past few decades, with the increase in industrialization and pressure of increasing human activity along the coast zone, waste disposal has become a serious problem. These waste materials comprise high volume of sewage sludge and offshore installation, to low volume but toxic waste such as toxic chemicals and even radioactive material.

The world population is 6 billion and counting. It is predicted to reach 10 billion by the end of the twenty-first century (Merrick, 1990). This will correspondingly increase industrial activity along the coastal zones with the related increase in domestic waste. It will be even more so in the growing economies such as India, China, Indonesia, and others. Therefore, it is time to think of ecological services to reduce the waste, management of waste practices, and degree of protection for oceanic environment.

There shall be definite need to minimize waste production, maximize waste recycling, and, of course, more of continued use of lands itself for waste disposal. Unfortunately, some of the waste types have already been dumped in the deep sea. Other waste types are now being discussed and planned to contain damage.

Our interest here is to examine the potential mining of the deep sea, incorporating (a) the nature of waste, (b) their transport and disposal methods available offshore, (c) the expected impacts on the marine environmental, and finally (d) what is the direction needed to assess the acceptability of the waste disposal in the deep sea. Other types of waste whose study is needed comprise radioactive wastes, waste generated due to offshore installation/sewage sledge, dredge spoils, and lastly carbon dioxide.

11.6 Indian Ocean and Climate Change

Even as the world tries to make sense of decisions taken at the recent climate change confer-ence in Paris, scientists have come up with fresh evidence of how global warming is begin-ning to nibble the food chain right in India's neighborhood, the Indian Ocean. Oceans play a critical role in both short- and long-term weather and climatic patterns. Nearly 90% of extra heat generated due to emission of greenhouse gases from the landmass is absorbed by oceans, warming them up. Indian Ocean, considered one of the most productive seas, has seen warming greater than other oceans. The warming in Indian Ocean during the past century has been estimated up to 1.2°C, which is a matter of concern, as it is very large compared to a global surface warming of up to 0.8° C during the same period.

Studies conducted by Indian Institute of Tropical Meteorology (IITM) have shown that the warming of Indian Ocean is affecting productivity of its marine ecosystem. This affects food plankton and other food necessary for fish growth and development and con-sequently for fish production. This, in turn, is resulting in dwindling fish catch rates in the Indian Ocean.

The decline in phytoplankton is a matter of concern since it is important food chain for fish growth. Phytoplankton contains chlorophyll and provides food for a range of sea creatures including fish. In the Indian Ocean, rapid warming is playing an important role in reducing the phytoplankton up to 20% during the past decades. Evaluation of data collected from satellite, in conjunction with climate models based on past climatic data, shows that the decline is up to 30% in the western Indian Ocean during the last 16 years. Observational field data from ocean profiling floats deployed in the Arabian Sea was used to validate satellite data.

Warmer ocean surface temperatures result in less dense water on the surface and denser water in the subsurface. This, scientists say, inhibits vertical mixing of nutrient-rich sub-surface waters with surface waters. Vertical mixing is necessary to bring nutrients into upper layers of the ocean where sufficient light is available for photosynthesis. Besides sustaining the marine food web, phytoplanktons absorb solar radiation and modulate the upper ocean heat flux. Thus, they have a major role in influencing various climate pro-cesses including the carbon cycle and sea surface temperature.

The abundance of tuna and other fishes in the Central Indian Ocean is related to link with phytoplankton availability and spread. The Food and Agriculture Organization (FAO) data indicate that the Indian Ocean accounts for 20% of the total tuna catch. Available data from the Indian Ocean Tuna Commission also shows that the tuna catch rates in the Indian Ocean have declined by 50%–90% during the past five decades. Much of this decline has to do with increased industrial fisheries, but reduced availability of phytoplankton is a major player. The studies of National Institute of Oceanography and satellite data prediction by the Indian National Centre for Ocean Information Services (INCOIS), Hyderabad, have contributed to our knowledge and understanding of this phenomenon.

Weakening of southwest monsoon is a matter of concern, because the food production is related to rainfall. The land–sea surface temperature difference over South Asia has been decreasing due to rapid warming in the Indian Ocean and a relatively subdued warm-ing and even cooling in some parts over the subcontinent. This enhanced warming of the ocean reduces land–sea temperature difference, dampening monsoon circulation, and affecting rainfall. It indicates (a) lowering of phytoplankton to about 20% in the Indian Ocean during the past few decades, and (b) lowering of marine productivity could be attributed to rapid warming in the Indian Ocean, and (c) future climate projections indicate further warming and subsequent reduction in marine productivity.

References

Amann, H., 1985. Development of ocean mining in the red sea. *Mar. Min.*, 5(7): 103–116.

Amos, A. F., Garside, C., Gerard, R. D., Lavitus, S., Malone, T. C., Paul, A. Z., and Roels, O. A., 1973. Study of impact of manganese nodule mining on the sea bed and water column. In: *Inter University Research on Ferromanganese Deposit on the Ocean Floor*. Department of Environment, National Science Foundation, Washington, pp. 221–264, 358.

Amos, A. F. and Roels, O. A., 1977. Environmental aspects of manganese nodule mining. *Mar. Pol. Int. Bull.*, 18: 160–162.

Baturin, G. N., 1982. *Phosphorites on the Sea Floor*. Elsevier, Amsterdam, p. 343.

Birch, G. F., 1980. A model of penecontomporaneous phosphatization by diagenetic and authigenic mechanisms from the western margin of southern Africa. In: Bentor, Y. K. (ed). *Marine Phosphoritus—Geochemistry Resource Genesis*. Vol. 29, Soc. Economic Paleontologists and mineralogist. Spl publication, Wisconsin, pp. 18–33.

Bremner, J.M., 1980. Concretionary phosphorate for SW Africa. *J. Geol. Soc.*, London, 37: 773–786.

Bischoff, J.L. and Piper, D.Z. (eds). 1979. Marine Geology and Oceanography of the Pacific Nodule Province. Plenum, New York, 842 pp.

Cook, 1988. Eustatic sea level changes, environmental, tectonics and resources. Brunn Memorial Lecture. IOC, UNESCO, 34, 35–57.

Cruickshank, M. J., 1990. Mining technology for Gorda Ridge sulfide. In: G. R. McMurry (ed.). *Gorda Ridge—A seafloor spreading center in the United States Exclusive Economic Zone*, Springer, New York, pp. 211–221.

Degens, E. T. and Ross, D. A., 1969. *Hot Brines and recent Heavy metal deposits in the Red Sea*. Springer-Verlag, New York, 600.

Glasby, G. P. and Roonwal, G. S., 1997. Marine pollution in India: An emerging problem. *Curr. Sci.*, 68: 495–497.

Merrick, T. W., 1990. US population a consider priority for the 1990s. Population Reference Bureau, 20.

Murray, J. and Renard, A., 1891. Deep-sea deposits. In: *Rep Sci Rep Explor Voyages HMS Challenger*. The Royal Society, London, pp. 521.

Nawab, Z., 1984. Red sea mining: A new era. *Deep. Sea. Res.*, 31(A): 813–822.

Rona, P. A., 1983. Exploration for hydrothermal mineral deposits at sea floor spreading centres. *Mar. Min.*, 4: 7–38.

Roonwal, G. S., 1986. *The Indian Ocean: Exploitable Mineral and Petroleum Resources*, Vol. XVI, Springer-Verlag, Heidelberg, pp. 198.

Sakasegawa, T. and Matsumoto, K., 1997. Deep sea manganese nodule mining. Marine Industrial Technology, UNIDO, 1–18.

Sharma, R., 2001. Indian Deep Sea Environment Experiment (INDEX): A study for the environmental impact of deep sea bed mining in the Central Indian Ocean. *Deep. Sea. Res. II*, 48: 2395–2426.

Sharma, R., 2005. Indian Deep Sea Environment Experiment (INDEX): Monitoring the restoration of marine environment after artificial disturbance to simulate deep sea mining in the Central Indian Basin. *Mar. Georesour, Geotech.*, 23: 253–427.

Sharma, R., 2011. Deep sea mining, economic, technical, technological and environmental consideration for sustainable development. *J. Mar. Technol.*, 45: 28–41.

Sharma, R, Nath, B. N., Parthiban, G. and Sanker, S. J., 2001. Sediment re-distribution during simulation benthic disturbance and its implication on deep seabed mining. *Deepsea Res.II*, 48: 3363–3380.

Thiel, H., 1991. From MESEDA to DISCOL—a new approach to deepsea mining risk assessment. *Marine Mining*, 10: 369–386.

Thiel, H., 1992. Deep sea environmental disturbance and recovery potential. *Intern. Rev. Gas. Hydrobiol.*, 77: 331–339.

Thiel, H. Bluhm, Borowsky, C., Bussau, C., Gooday, C., Mayburg, A. J., and Schriever, C., 1992. The DISIOL Project (The impact of mining on deep sea organism). *Ocean Challenges*, 3: 40–46.

Thiel, H., Schriever, G., Bussau, C. and Borowski, C., 1993. Manganese nodules crevice fauna. *Deepsea Res.*, 40: 419–423.

12

The Future Options: The Growing Ocean Economy

12.1 General

The UN International Law of the Sea Agreement (UNCLOS) came into effect on November 16, 1994. This is the final outcome of a decade of ongoing negotiation and disputes, after more than 168 countries, including India, accepted it. The UNCLOS comes under the competence of the UN General Assembly. It gives 200-nautical-mile exclusive economic zone (EEZ) to the nations. Within the Indian Ocean (Figure 12.1), one can see the change, and consequently responsibility of the states for a sustainable development goal (SDG) aim to future activity. They have the responsibility for the conservation, management, and use of a large sea area around the mainland and island territory.

How to manage this added responsibility for an SDG goal? To be able to manage EEZ zone, society needs to undertake and understand long- and short-term environmental problems (Chapter 11) faced by various sectors of the zone and consequences associated with utilization of mineral resource and future developments, in the use of both EEZ and activity in the open sea. This would call for an understanding of their likely environmental impact on the assessment of their potential clean technology and sound environmental management practice (EIA, Chapter 11), highlighting the economic, social, political, technological, and legal barriers associated with the development of coherent management policy for ocean utilization. For these groups belonging to public and private sector undertakings involved in the discharge of sewage and industrial wastes, extraction of hydrocarbon, and hard minerals from both deep and shallow water, deep-sea fishing, recreation, and tourism all are necessary aspects that need to be understood. In Chapters 9, 10, and 11, an attempt has been made to highlight issues for an inclusive development to attain SDG.

The United Nations data estimates the present world population at approximately 5.8 billion. It is expected to reach about 9.8 billion by the year 2050. More specifically, the population would increase by 75 million people per year over the next 50 years. The current annual global consumption of resources consists of approximately 3.5 billion tons of coal, 3 billion tons of petroleum, 650 million tons of iron and steel, 20 million tons of aluminum, 12 million tons of copper, 900,000 tons of nickel, and about 25,000 tons of cobalt. Because of availability and consumption of these resources, the people live conveniently and comfortably. It is generally considered that the world population will continue to increase and, for most, living standards will also continue to improve. This is to say that society is going to be faced with an even greater consumption of resources than ever before. These resources exist not only on land but also on the deep seabed, for example, manganese nodules, cobalt-rich manganese crusts, hydrothermal polymetallic sulfide deposits, phosphate nodules, heavy minerals, or beach sand for their rare earth elements (REEs) and so forth.

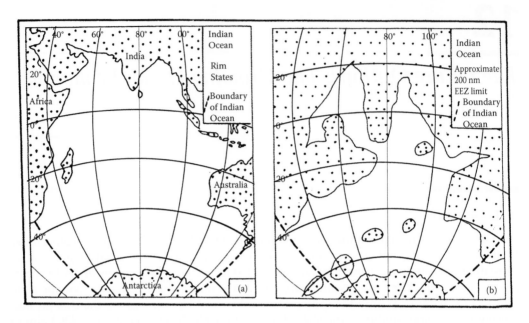

FIGURE 12.1
EEZ in the Indian Ocean. (a) The ocean without EEZ; (b) the ocean with EEZ. (From Qasim, S. Z. and Roonwal, G. S., 1998. The Exclusive Economic Zone. In: *Living Response of India's Exclusive Economic Zone*. Omega, pp. 1–4, 140 pp.)

Society needs to recognize these resources on the land and on the deep seabed as earth's resources. For the time being, society shall possibly be able to continue depend on land resources. However, since the consumption of resources continues to increase, it is anticipated that in the near future, world need to mine resources from the deep seabed in order to maintain the standard of living to which people have become accustomed.

Society receives great benefits from marine resources. Major marine resources include biological, energy, mineral, and material resources. Yet, society needs to only enjoy the benefits of these resources now, but also take responsibility for maintaining these natural resources for generations to come. The United Nations Convention on the Law of the Sea (UNCLOS), Part XI, declares that the deep sea and its resources are the common heritage of humankind. With this in mind, it is necessary and vital to collect scientific data and information on present sea conditions, as well as on biological, energy, and mineral resources for their sustainable development (ISA, 2001, 2012).

These days, environmental issues, such as global warming, desertification, and so on, have become the center of public debate, and these issues will become more important in the years to follow; thus the twenty-first century will be the environmental century. Because the consumption of resources would continue to increase, as a consequence, the global environment will be affected. Therefore, mutual cooperation to secure a comfortable living standard while sustaining the global environment is imperative. To achieve this, it is important to establish a new socioeconomic order for the recycling of resources. Hence, "sustainable development" goal is likely to be the most important issue of the twenty-first century. This chapter therefore focuses on processing the deep seabed minerals as a future raw material source and very briefly introduces exploration, mining, and processing total system of deep-sea mining and environmental research on them as well as energy and living resources from the ocean.

12.2 Exclusive Economic Zone in the Indian Ocean

The UNCLOS, which came into force in November 1994, has created a new regime for the oceans by giving all maritime nations exclusive rights over economic activities in a region that is now termed the EEZ. It extends into the ocean up to 200 nautical miles from every point on the sea waters, including island territory. Figure 12.1(a) shows both the configurations of islands and continents in the Indian Ocean and (b) after the EEZ regime, how much of the "free" areas would be available in the Indian Ocean. India's EEZ is about 2.02 million km^2, which is nearly to 18.7% of its land area. In a rapidly changing scenario, the utilization of the EEZ, and the rest of the ocean have to be looked into with considerable foresight responsibility and accountability (Laughton, 1996). The EEZ is also important since it includes the narrow coastal zone where most of the economic activity takes place, which often influences the ecological balance of the coastal water and consequently economic activity like fishing in the coastal area (Glasby and Roonwal, 1997). Outside the EEZ, only the deep-sea minerals of the seabed come under the control of the International Seabed Authority (ISA), and not the mobile resources such as fisheries. This is why the EU–Canadian fishing dispute of 1994–1995 had to be amicably resolved bilaterally rather than by the ISA.

According to the UNCLOS convention, when the continental shelf gets established or extended beyond 200 nautical miles, it may also claim seabed biota, minerals, and hydrocarbon resources, upto nearly as much as 350 nautical miles or 684 km from the territorial sea base line, or say additional 100 nautical miles (185 km) beyond the 2500 m isobaths. Again, where submarine elevation occurs as a natural component of the continental margin such as plateau, rises, banks, or spurs, the limit could be pushed even beyond 350 nautical miles.

This new regime would lead to far-reaching consequences for the conduct of ocean research, since the management of the marine resource is the responsibility of the states. As a result, fishing within the EEZ of one nation by another nation would amount to infringing upon the exclusive rights of the nation, unless the activity becomes a part of a joint enterprise or has a formal clearance from the nation under an agreement sharing resources. Yet another aspect, which emerges from EEZ regime, is that some of the small island nations with a small population and land now have a very large EEZ. Since such nations comprise several islands, whether they are in habitat or not, the EEZ is measured from each island. The Indian Ocean has several such examples. It is often felt that real beneficiary of the UNCLOS regime is the developing and island nations. Most of them have declared their EEZ soon often ratifying UNCLOS. The developed nations, such as the United States and the UK, have not yet officially announced their EEZ.

12.3 Implications of EEZ in the Indian Ocean

The Indian Ocean, which has long remained poorly studied, even on global fisheries, has now evolved into an important fishing zone after the EEZ has come into force. The average annual marine fish catch in the Indian Ocean is low compared to the global fish catch. Because of intervention of FAO, the Indian Ocean Tuna Commission has been promoted. Impressed by the APEC, ASEAN, and ASCAN Regional Forum, cooperation among the Indian Ocean Rim (IOR) nations has increased. It includes Mauritius, India, Australia, South Africa, Kenya, Singapore, and Oman in 1996, but has since now strengthened more.

This was followed by a conference on "The Making of an Indian Ocean Community," which was an activity point of better cooperation and new vista opened for all sectors in the IOR. One looks ahead to 2020 and further to see development in the UNCLOS and the state practice relating to management and conservation of fishery resources in the EEZ of the Indian Ocean regulation. Keeping this aspect and the challenge faced, the following suggestions are worth considering by all IOR nations.

1. National ocean fishery policy to include and integrate coastal area management (ICAM), island development, and ecofriendly fishing community fishing to protect the interests of the local fishing community. Use of modern methods for remote sensing and training service and review prospect.
2. Create regional cooperation in promoting sustainable fishing (part of SDG).
3. Regional cooperation in assessing marine fishing potential of the Indian Ocean for high-sea fishing also.

12.4 Economics of Deep Sea Mining

Great excitement was generated when the manganese nodule resources were discovered and their resultant addition to world metal supply (Mero, 1965). Later, the acceptance of UNCLOS was another significant event. This led to the creation of the ISA. Now, it is accepted that besides nodules, there is interest in cobalt-rich crusts and seafloor sulfides, and hydrocarbon and gas hydrates can be added to the list. Like any mining activity, in case of deep-sea mining, the environment impact assessment is required. Therefore, at present, there is an emphasis as much as on mining technology as on the environment assessment that is likely to happen as a sequel to deep-sea mining. In this respect "Use and Misuse of the Sea Floor" (Hsu and Thiede, 1992) is an important guideline. Undoubtedly, the technology needed for deep-seabed mining has benefited from the experience gained though offshore deepwater hydrocarbon recovery. At present companies and consortia are engaged in refining the technology, focusing on arising issues and other aspects.

These issues have been elaborated in Chapters 9 and 10. What needs to be examined is the present situation and whether deep-seabed mining is profitable. There are several aspects of it—the role of ISA's Enterprise, the technology transfer, the relinquishment of an equal area of for the Enterprise and even how to mine, how to share expenditure, cost, and benefit, or profit, which is likely to be generated. Several models are available for nodules such as the well-known MIT model. The need for the better understanding of the several aspects of deep-seabed mining also got a push by the application of Papua New Guinea for sulfide exploration and eventual mining in Manus Basin, southwest Pacific Ocean. Cruickshank (2002) has brought out different aspects of the seabed mining, such as

1. Within UNCLOS–ISA–baseline information and delineation of mine lease.
2. The nations' need to respect UNCLOS and adopt it, and details regarding this.
3. Resolution of difference and disputes redressal within UNCLOS.
4. Promote science research in ocean sector.
5. Rights of coastal states and island nations.

The concept of EEZ has changed the situation. Figure 12.1 (a and b) shows land and ocean in the Indian Ocean. With a 200-nautical-mile EEZ limit, how would the Indian Ocean look can be clearly seen from the figure. When and if the extension of continental shelf areas is announced, as it is likely, this would leave very little of free ocean for any work study and all aspects because the EEZ extension would squeeze the free space from the international waters. Therefore its execution can succeed only with cooperation, mutual trust, and respect. Glasby (2002) has also deliberated on past and future prospects in deep-sea mining. He focused attention on heavy subsidy, whether by government or consortia supported by government, for the deep-sea mining activity. Within the new phase in UNCLOS, India, China, Japan, and Korea have committed to a long-term program in the Indian Ocean. This is quite in line as recommended in the desk study of the marine science and technology—the environmental risk of the European Union 1997.

Though efforts are on for developing technology seafloor nodule mining, economic analysis does not give a promising support for this. The metal values and market prices of metals do not make it as an attractive venture at the present time. However, the seabed massive sulfide with their higher metal values of Zn, Cu, Pb, Au, and Ag make it more attractive. Although as compared to land-based deposit such as Kurukoo in Japan or in Cyprus, the size of the SMS on the seafloor is small—about a million tonnes per location. Yet when looked at a cluster of such locations, it would certainly be a very promising venture. This would be economically viable and ready to face metal market fluctuation.

Starting with the Red Sea metalliferous mud, the Atlantis II deep in the Central Red Sea has been studied and commercial assessment conducted (Backer 1980, Blissenbach and Nawab 1982, Herzig, 1999; Scholten et al. 2000). Further details of SMS are available in Rona and Scott (1993), and Scott (2008), Rona (2008). Southwest Pacific's East Manus Basin is now receiving attention from private consortia, Nautilus, and could be the first SMS mining venture to begin. On the optimistic side, plans to make seabed mining a reality are in progress, with Canadian Company Nautilus Minerals expected to initiate the Solaware-I project in Papua New Guinea, the first of its kind, by early 2019. The project will yield copper and gold deposits mined from the seafloor. A report suggests that seabed copper mining would be more sustainable than terrestrial mining. In the Indian Ocean (Chapter 4), China, Germany, India, and South Korea have exploration licenses given by the ISA. The Triple junction and the SWIR are more attractive sites for it. From the above discussion, an outlook for mining of deep-sea nodules and cobalt-rich crusts look somewhat uncertain in the present situation. The SMS deposit has overtaken nodules in terms of economics and feasibility; only time would let it be known how early SMS deposit would be mined.

12.5 Estimates of Demand

The average SMS deposit contains up to 20 different metals in a variety of combinations. Iron is ubiquitous but will not be considered in this study as huge reserves exist on land, and it will not be economic to extract from the sea in the foreseeable future. The most economically interesting metals found in the samples so far collected are zinc, copper, silver, and cobalt. The market for each of these metals will be considered in turn with a view to assessing their likely future demand.

12.5.1 Zinc

The main use of zinc is still that for galvanizing iron to protect it from oxidation, which accounts for 40% of all its consumption. A further 32% of production is used in die castings, and 14% is employed to make brass fittings and decorations. The remainder finds its way into paints, rubber, and medicines as zinc compounds. Finally, a new, growing application for zinc is in nickel–zinc batteries. Large amounts are used annually and zinc is the sixth in the table of metal consumption. Its main attributes are resistance to corrosion, ease of molding, and affordable prices.

The main ore for zinc is sphalerite (ZnS). It is an extremely common mineral often found in association with the ore for lead, galena. Known reserves are large, and at the present rate of growth in consumption (almost static), reserves will stretch into the next century. Zinc production is still rising slowly, though the price is currently stable. The prospects for the zinc market are very closely linked to steel production, due to galvanising. It follows that demand is likely to remain fairly constant with some growth possible due to demand expansion in the developing economies. Increased demand is likely should nickel–zinc batteries become more widely used, such as in electric cars. It is unlikely that new substitutes will be developed for zinc due to its low cost. However, if the price begins to rise due to exploitation of lower grade ores, substitution may become a possibility.

12.5.2 Copper

More copper is used annually than zinc. Its biggest use is in electrical appliances; transformer and generator windings, switchgear, etc., and in the ubiquitous copper wire. Copper's second most important application is in the construction industry, primarily for plumbing. It is also widely used in heat exchangers, for example, the car radiator. Other uses of copper include bearings and bushes, gas and oil lines, coinage, jewellery, etc. The most useful properties of copper are its high electrical conductivity and extreme ductility.

Copper is a common metal, occurring in many minerals. The most important ores of copper are chalcopyrite ($CuFeS_2$)/bornite (CuS, FeS_4), and chalcocite (Cu_2S), the first providing by far the most copper. The ores are found in a variety of geological settings, one of the most important in recent years being "porphyry deposits." These are disseminated copper ore bodies formed in underground magma chambers, which have cooled rapidly. They are very large and low-grade (typically less than 0.5% copper). Consequently mining is extremely energy intensive.

Copper is one of the earliest metals used by man. Its current value in electrical wiring and other engineering materials is well known and there are multiple applications associated with its ability to conduct electricity and heat, plasticity, and resistance to corrosion. The major portion of copper produced in the world is utilized in the electrical industries. Table 12.1 shows the world resources of copper, from which it is indeed clear that Indian copper resources are not large. Considering the population and demand, India is very deficient in copper, so much so that even for the electrical industry and cable wires, aluminum is utilized as a substitute for copper, which has over the years resulted in severe fires in some commercial complexes due to short circuiting. As mentioned above, Indian copper reserves are about 1% of the world resources (Table 12.2), but the grade is rather low, varying between 0.91% and 2.0%, with the bulk of the deposit below 1.5% in copper content. India imports the bulk of her copper (between 70% and 85%), so much so that even the production of ores from existing mines falls short of the smelting capacity and is met

TABLE 12.1

World Reserves of Copper

Country	Quantity (1000's Short Tons)	% of World Total
World	503,000	
United States	93,000	18.49
Australia	8000	1.59
Canada	34,000	6.76
Chile	93,000	18.49
Papua New Guinea	10,000	1.99
Peru	35,000	6.96
Philippines	19,000	3.78
South Africa	3000	0.60
Zaire	28,000	5.57
Zambia	32,000	6.36
Other Market Economy Countries	82,000	16.3
Poland	14,000	2.78
USSR	40,000	7.95
Other Central Economy Countries	12,000	2.39
India	5000	0.99

Source: Indian Mineral Yearbook, 2012, Indian Bureau of Mines, Govt. of India, Nagpur.

by importing copper concentrates to be able to run the smelters. Furthermore, the cost of production is very high.

12.5.3 Silver

One of the most important uses of silver is in jewellery. Indeed one of the largest known reserves is in India, where it has been made into jewellery and accumulated for centuries (estimates range from 4 to 7 billion ounces). The ornamental use of silver is now rivalled

TABLE 12.2

World Cobalt Reserves (Metal Content)

Country	Short Tons	% World Total
World	1,600,000	
Australia	54,000	3.38
Botswana	29,000	1.81
Canada	33,000	2.06
Finland	20,000	1.25
Morocco	14,000	0.88
New Caledonia	300,000	18.75
Philippines	210,000	13.13
Zaire	500,000	31.25
Zambia	125,000	7.81
USSR	350,000	21.88

Source: Indian Mineral Yearbook, 2012, Indian Bureau of Mines, Govt. of India, Nagpur.

by its use in the form of silver halide in photography. Silver is also being increasingly used in the electronics industry due to its malleability and high conductivity. It is also widely used as a catalyst.

The most common form of silver ore is argentite (Ag_2S), but it also occurs in its native form. Silver is usually produced as a by-product of other mining operations, usually for copper, lead, zinc, or gold. Any operation to mine SMS is likely to produce a large proportion of its revenue from silver, given the present world market. Indeed, the Red Sea project showed that silver is the most profitable part of the whole operation.

12.5.4 Cobalt

The most important use of cobalt is in the production of super alloys, primarily used in high-speed jet engines and spacecraft. The second major use is in permanent magnets, where it produces stronger magnets than iron. Cobalt is also used in a variety of chemical applications, such as glazing, enamels, and paints.

It is apparent from the above reviews that demand for the major constituent metals of SMS deposits are fairly static and that adequate reserves (for short to medium term demands) are known to exist on land. In the long term, the prospects for deep-sea mining look far more promising as available land-based alternatives became progressively lower in grade and consequently increasingly expensive to extract. Metals markets are, however, notoriously unpredictable and any large-scale price rises in the relevant metals would increase the SMS as a resource. It is also important to bear in mind that so little is known of the average composition of SMS that future finds could radically alter current thinking on which metals are the most desirable commodity. Finally, it is possible that strategic considerations may lead to the mining of SMS on a nonprofit basis by a government wishing to ensure a secure supply of strategic minerals such as cobalt.

The magnetic properties of cobalt give it many engineering uses. It is ferromagnetic at room temperature and has the highest Curie point. Nearly one quarter of the world production of cobalt is used for magnets. Cobalt indeed has other uses in the formation of alloys with different metals. High-temperature steel alloys are used within turbines, space vehicles, petrochemical equipment, power-generating equipment, etc., because cobalt alloys have a high resistance to thermal fatigue and high melting points. World reserves of cobalt show India's total deficiency in cobalt, although it is a very necessary metal. India is totally dependent on imported cobalt, the use of which since industrialization has increased dramatically.

12.5.5 Nickel

Modern industry utilizes a great deal of nickel, and nearly half of the world production is used in the form of alloys of iron to make different types of steel, to which it adds strength. Therefore nickel is incorporated into automobiles, locomotives, ships, airplanes, agricultural machinery, machine tools, power generating equipment, mining machinery, to mention a few examples. World nickel resources (Table 12.3) clearly show that India has almost no resources of nickel, though there are some indicated reserves.

Thus, India has to depend completely on imports for nickel, from sources such as Canada, Australia, and Russia. Furthermore, of total reserves (Table 12.4) nearly one half is estimated to contain 0.8% nickel. However, nickel consumption in India has grown at a rate of slightly over 10% per annum.

TABLE 12.3

World Reserves of Nickel Ore

Country	Reserve (Short Tons)	Grade % Nickel	App. % World Reserves
World	60,000,000		
United States	200,000	0.8–1.3	0.33
Canada	8,600,000	1.5–3.0	14.33
New Caledonia	15,000,000	1.0–3.0	25
West Europe	28,000,000	0.2–4.0	46.67
Cuba	3,400,000	1.4	5.67
East Europe/USSR	4,800,000	4.0	8
India	137,000		

Source: Indian Mineral Yearbook, 2012, Indian Bureau of Mines, Govt. of India, Nagpur.

12.5.6 Manganese

Manganese is an essential component of steel. About 90% of world manganese output is used as ferro-alloys by the metal industry in the manufacture of steel. Pure manganese is used as a purifying agent. Although India has moderate reserves of manganese (Table 12.4), good quality manganese is being rapidly depleted.

12.6 Future Demand and Ocean Economy of Deep-Sea Mining

With a population growing at a rate of 1.08% and industrial growth at 7%, even up to 8% in some sectors, India offers great new markets. However, it also has an urgent need to process raw materials. One can see no difficulty in meeting the requirements of the synthetic, fiber, and other polymer products, also possibly food supplies, if the monsoon is regular and good. But as regards base metals and even some other metals such as high-quality manganese ores, there is obviously serious difficulty. The following is attempted to analyze what could be the future demands for these metals.

TABLE 12.4

World Reserves of Manganese Ore

Countries	Reserves ('000 st)	% World Total
World	6,000,000	
Australia	300,000	5
Brazil	95,000	1.5
Gabon	165,000	2.8
India	65,000	1.1
South Africa	2,200,000	36.7
EC Countries	59,000	1.0
CIS	3,000,000	50.0

Source: Indian Mineral Yearbook, 2012, Indian Bureau of Mines, Govt. of India, Nagpur.

The issues relevant to get a comprehensive benefit of the deep-sea mining are (a) the competitiveness of seabed mining as compared with land-based mining, (b) the potential impact of mineral supply from seabed on the world market for the minerals, and finally (c) the income effects that can be expected from deep-sea mining. The empirical analysis not only improves understanding of the economics of deep sea mining, but also stimulates further research in the area.

Since 1876, when HMS *Challenger* discovered manganese nodules on the ocean floor, much has been discussed and written in an attempt to understand how deep-sea mining would change world order. Primarily the deep-sea mining will compete with land-based mining for metals such as nickel, and cobalt, copper, zinc, gold, and silver. Apart from nickel, the manganese nodules contain other metals of which cobalt, copper, and manganese are significant in economic terms. Though from the point of view of revenue, nickel is the most important component so that deep-sea mining can best be compared with nickel mining on land. As is known, cobalt is usually mined as a by-product of nickel and copper, and thus the analysis can be restricted to nickel without significant loss. The seafloor massive sulfides (SMS) contain all these elements.

New land-based nickel mining operators are based mostly on lateritic ore deposits. Thus, one can make a comparison of the cost between deep-sea mining and land-based mining, which implies comparing the cost of recovering and smelting of nodules with those on land-based lateritic nickel ores. Similar calculation can be done for SMS deposits (Rona, 1988; Chung, 2003).

A new dimension is added to this with the discovery in 1979 of the seafloor massive sulfide (SMS) on the midoceanic ridge system, particularly in the Pacific (EPR), the Atlantic (MAR), and the South Pacific subduction zones, and more recently is Mid-Indian Ocean Ridge. These deposits are based on hydrothermal activity generated by the precipitation of rich ores of copper, zinc, and others, but no nickel in them. Therefore, the two calculations on economics of deep-sea mining are different. At present, estimates are derivable on and include the manganese nodules parameters, involving (a) an estimation and actual value of the fixed and variable cost and (b) the trend shown by these costs (Black, 1980).

Fixed and variable costs of deep-sea mining have nodules available from sources such as most discussed Massachusetts Institute of Technology (MIT) model. These MIT model provides a cost estimate for the areas of research and development, prospecting and exploration, mining, transportation, processing and smelting of manganese nodules in a hypothetical deep-sea mining operation. In addition, the MIT study presents estimate of revenues from nickel, copper, and cobalt. As study result shows, it has achieved more accurate results with time. However, because of the complexity of the deep-sea mining project study and different technologies and methods considered for mining operation, it restricted to generalization. Therefore, only the parameters are considered rather than actual assessment of the cost of deep-sea mining at this stage. The parameters are

1. Size of the plant (annual capacity)
2. Recovering methods (mining of nodules)
3. Smelting methods (processing of nodules) to extract the metals.

Similar to land mining operation, the deep-sea mining is made up of fixed costs for areas of research and development, prospecting and exploration, mining transportation, processing and smelting, and variable costs for the areas of mining, transportation, processing, and smelting and assumed quantity of metals recovered. Because the metal price stocks are greatly variable, any costing has to be conducted according to the prevailing local conditions and the likely profit generation.

In order to get an idea of the benefit of deep-sea mining, one needs to examine it in relation to land mining. Since the deep-sea mining is yet to commence, no actual figures exist. Therefore land mining experiences and lessons are extrapolated to work out the cost analysis of the future deep-sea mining projects. As the land mining project costing depends on several factors of demand and supply, the technology, wages, transport, equipment, and others, so would be the situation in the case of deep-sea mining.

One study proposed by Black (1980) gave detailed data on factor input in deep-sea mining. This is based on the MIT model. Details of the study included input of chemicals, lime, electrodes, and energy (heating oil, coal, lime, electrodes, electricity, etc.) (One has to correlate the nickel equipment with the energy concession balance.)

It is also suggested that the prices of materials and energy in land-based mining are similar to these of the deep-sea mining, which is another aspect that needs special mention. Since most of the nickel mining is being carried out in developing countries where nickel mines exist and where the cost of living and wages are very low, the wages in these mines is low. But the commonality is that both types of mining shall be energy intensive. The deep-sea mining shall be possibly more material-intensive than land-based mining. In this respect, considering the displacement and rehabilitation cost, the capital-related costs of the land mining appear to be on the higher side than those are all variable factors. Today the cost of environmental restoration and mine closure costs are increasing.

This present scene is changing rapidly: one would expect more contribution from alternate energy sources and substitutes are increasing. Therefore the costs shown are only indicative. The detailed costing has to be worked out along with a detailed project report. But the indicator is not going to change substantially because the wages are likely to increase. So final analysis will be based on changing pattern by demand on whose strength supply and therefore mining operation will depend (MIT Report 1978).

12.6.1 Economic Impact of Deep-Sea Mining

When deep sea-mining begins, there shall be additional minerals supply from the seabed. This would result in both short-term and long-term impact on the supply, production, and consumer surpluses. The other likely impact would be the revenue of ISA generated through fees and royalty paid by mining companies. The ISA is now discussing finer aspects of revenue sharing in the event of future deep-sea mining projects. In addition to these direct effects, one would also expect a few indirect effects of the development in technology and its interindustry linkages, before actual transfer of technology on a "soft" basis to the Enterprises wing of ISA. To go into finer details, one needs to understand the full impact of seabed mining on consumption and production surplus. The issue of ISA revenue from seabed mining needs fine-tuning considering the net gain for major products and consumers, as contained in edited volume by Donges (1985).

The Indian Ocean countries are mostly developing economies, with high density of population. The social and government plan to support the development projects to uplift the quality of people would require a huge sum of money, expertise, and planning for implementation. Therefore, mutual cooperation in the region would be an important factor in this goal. Then only the economic benefit of the "ocean economy" would support and provide funding from income to reach the SDG.

Projection of demand for nonferrous mines in the Indian Ocean context requires consideration of a variety of factors including

1. demand projection
2. industrial demand, supply, and reserve position
3. government resources, since metal mining is state-owned
4. availability of and cost of energy
5. appropriate technology
6. raw materials
7. environmental factors
8. substitute availability

12.6.2 Possible Mining Technology

The available technology for the ongoing dredging of heavy mineral sands in the offshore areas of Malaysia, Thailand, and Indonesia are well known. With the advancement of engineering technology, modifications will be adopted to optimize the dredging process and reduce the time as well as the cost of operation. Likewise, in the onshore or nearshore areas of Sri Lanka and India, heavy minerals are mined. The offshore exploration and exploitation of hydrocarbons in the relatively shallow water offshore zones is also being practiced. With the determination and need to locate and produce more hydrocarbons, the engineering and achievement of the technology has been to go the higher water depth of up to 400 m. Still at the stage of research and experimentation are the technologies for future marine resources namely:

1. Manganese nodules, as found in the Central Pacific or the Central Indian Ocean Basin
2. Metalliferous muds as found in the Red Sea
3. Massive sulfides from the Mid-Oceanic Ridges and subduction zones and cobalt-rich ferromanganese crusts.

Despite the uncertainty surrounding the Law of the Sea Treaty, several consortia not only prospected and explored for nodules, but also refined their technology with the objective of commercially mining them. Simultaneously, the ore beneficiation and metallurgical techniques were perfected. Likewise, similar studies are being conducted on SMS.

Various international consortia have not only shown willingness to develop deep-sea mining but have also invested in and perfected to a large extent the mining technology. The technology includes nodule collector systems and attempts to simultaneously process the nodules to increase the metal values onboard.

The nodule collector systems include the continuous line bucket collector (CLB), which, in principle, is dredging with a very strong rope cable to withstand 4 to 5 km depth of the ocean. The buckets would be dragged on the seafloor along a loop as explained in Chapter 10. Experiments have been done with a single ship or two ships to ensure wider and more efficient coverage of the seabed. There are technical difficulties with this system; in particular one cannot be sure of the total nodules that would be collected by the system. But this is a relatively inexpensive method.

The other systems involve collection of the nodules via a tractor with a suction pump to suck up the nodules as well as a large net to collect the nodules (Chapter 10). The aim in each case is to collect at least 10,000 tons of nodules per day. Such a "hydrosweep" system may eventually be adopted when nodule mining begins. There are related questions pertaining to the stability of the ship and the concentration of the nodule material on the ship itself before transporting the slurry to the mainland. But above all is the question of the environmental effects that the disturbance of sediments in the water column would cause. Several studies in this direction are in progress, particularly in the United States and Germany.

The technology for the future mining of SMS deposits and the Red Sea type metal-rich mud is different from the technology proposed for nodule mining, though, in principle, the problem is still to haul the bottom metal-rich sediments to the ship as in the case of nodules. But given the very different nature of these solid rocks associated with black smokers, the metal-rich mud requires a different methodology. A study of the Red Sea material has shown that mining is technically feasible. Figure 10.4 in Chapter 10 shows schematic view of the mining systems. A discussion on the engineering aspects of this system can be seen in the various publications on the topic, a recent one being Amann (1989, 1990). For SMS mining, some details of study and results are available for South Pacific–Solwara deposits (Nautilus website).

For the mining of cobalt-rich ferromanganese crust, the planning for experimental mining is yet to happen. However, when such plans are made and executed, the main problem would be to prepare a dredging or cutting system that is able to scrape a thin layer of encrustation from the mainly basaltic substrate.

12.6.3 Factors Influencing Mining

The decision on whether or not to mine a particular marine resource is influenced by factors other than resource potential and extractability. Nearshore dredging of minerals has taken place for many years, but extraction in the high seas is a new phenomenon, and a consensus covering rights in international law has yet to be achieved. In the absence of such agreement, it is highly unlikely that any deep-sea mining will take place. Any mining operation is also strongly influenced by existing alternative sources, even more so when a radical new technology is required for exploitation as in the case of SMS. Therefore a thriving market in the relevant metal is essential for deep-sea extraction to become a reality. Finally, all modern mining operations must consider the environmental effects of projects. This is likely to be a crucial factor in any proposal to mine SMS, given the unique fauna normally associated with active vent emissions.

At present, two legal regimes exist governing a coastal state's rights regarding exploitation of seabed resources, one pertaining to the area within national jurisdiction, and the other covering activities in international waters.

12.6.3.1 Mining within National Jurisdiction

Two legal concepts can be invoked for the exercise of a coastal states' right to exploit all minerals within 200 nautical miles off its coast, the EEZ, and the continental shelf. The EEZ is a zone extending up to 200 miles from the baseline from which the breadth of the territorial sea is measured. In the EEZ, the coastal state has sovereign rights for the purpose of exploring and exploiting, conserving and managing the mineral resources of the seabed and subsoil. The EEZ is a concept of recent origin and has been embodied into the 1982

UNCLOS which has entered into force since November 1994. However, a large number of states have incorporated the EEZ into their national legislation and it can now be said to be part of international customary law.

The rights of a coastal state over minerals found on its continental shelf are governed by Article 1 of the 1958 Geneva Convention, which defines the continental shelf as "(a) the seabed and subsoil of the submarine areas adjacent to the coast but outside the area of the territorial sea to a depth of 200 meters, or, beyond that limit, to where the depth of the waters admits of the exploitation of the natural resources of the said areas; (b) to the seabed and subsoil of similar submarine areas adjacent to the coasts of islands."

This definition was somewhat vague in character and led to a series of disputes between coastal states. A more practical definition of a continental shelf was incorporated into the 1982 UNCLOS, which defines it as

> the seabed and subsoil of the submarine areas that extend beyond its territorial sea throughout the natural prolongation of its land territory to the outer edge of the continental margin, or to a distance of 200 nautical miles from the baselines from which the breadth of the territorial sea is measured where the outer edge of the continental margin does not extend up to that distance. (Article 76 para. 1)

Hence, the EEZ and continental shelf are synonymous in legal terms out to 200 miles, but the continental shelf, if it satisfies certain geological criteria, can extend jurisdiction beyond this up to a maximum of 350 nautical miles from baseline or 100 nautical miles from the 2500 m isobath. Another consequence of Article 76 is that areas of the deep sea (off the continental shelf) now come under a coastal state's jurisdiction if within 200 nautical miles from the territorial sea baseline. Seabed mining within national jurisdiction of a coastal state would be subject to the laws and regulations of that state but must not interfere with traditional freedom of the seas, such as the operation of ships and aircraft or the laying of submarine pipes and cables.

12.6.3.2 Mining in International Waters

According to UNCLOS, the seabed outside coastal state jurisdiction is termed "the Area" and accounts for 60% of the ocean floor. An international legal regime governing the exploration and exploitation in the Area of all "solid, liquid and gaseous mineral resources at or beneath the seabed" is elucidated by UNCLOS, which incorporates the following general principles:

1. The resources of the Area are the common heritage of mankind and as such are not susceptible of unilateral national appropriation.
2. Activity in the Area shall be conducted so as to promote the peaceful and rational development of the resources and shall be carried out for the benefit of mankind as a whole.
3. A new international organization called the ISA participates in the exploitation of the Area through its operating arm the Enterprise or through the granting of licenses for exploitation under its supervision by private contractors.
4. The relevant technologies used by private operators must be transferred to the Enterprise on fair and equable terms and conditions.

5. The minerals of the Area must be developed in such a way as to promote a just and stable price and to protect developing land producers against adverse effects. There is a mechanism of control for the volume and value of the metals produced from nodule mining but not for other types of deposits such as SMS.

Following preliminary exploration, the mining contractor makes an application to the ISA for an exploitation area divided into two parts of equal commercial value. The authority selects one part to be reserved for its own mining activity. The second part is subjected to a licensing system whereby the contractor is given exclusive rights for the exploitation of a mine site by the authority, in return for complying with a series of rules and regulations, i.e.,

1. The contractor is subject to production controls. He is obliged to transfer technology to the Enterprise (the operating arm of the ISA) on fair and reasonable commercial terms and conditions.

2. The contractor is subjected to financial terms involving the payment to the Authority of an administrative fee of $500,000 followed by an annual fee of $1 million.

Following the signing of the 1982 UNCLOS, the Preparatory Commission for the International Seabed Authority has been working on setting up the deep-sea mining regime and in particular on drafting a mining code.

Multilateral agreements such as the 1982 UNCLOS commonly take five to ten years to attract enough ratification (in this case 60) to enter into force. To avoid such delay and in view of the dissatisfaction of a number of Western nations with the mining provisions contained in the 1982 UNCLOS, a group of "like-minded states" have established a Reciprocating States Agreement (RSA or "minitreaty") to permit and regulate mining before the 1982 UNCLOS enters into force. This regime results from the formulation and adoption of similar national laws governing activities in the Area, which are coordinated by a series of international agreements aimed at reducing disputes between the participating countries.

This agreement has been presented by the participating states as an interim agreement pending implementation of the 1982 UNCLOS. Although it is designed to assure mutual recognition of participating countries' claims, it does not prevent states outside the agreement from claiming overlapping areas. The basic legality of this agreement is also challenged by the "Group of 77," a collection of over 100 nations who claim that the concept of deep-sea minerals as being the common heritage of mankind is already part of customary law, and as such unilateral claims are illegal. It is therefore likely that any attempt to mine the Area under national legislation would provoke stiff international opposition. Another problem with these national legislations is that they were drawn up with the exploitation of manganese nodules as their objective and so may not be directly applicable to other deep-sea deposits such as SMS.

In summary, the legality of mining SMS deposits must be viewed in two contexts. First, mining of deposits within 200 nautical miles of a coastal states' territorial sea baseline or within the continental shelf as defined in UNCLOS, could legally take place, subject only to the terms and conditions laid down by the relevant government. In contrast, mining of any minerals in the Area would be subject to significant legal uncertainties and would be open to challenge by third party states on the grounds that it violated international customary law. Mining of SMS in the Area under existing national legislations could also be challenged on the grounds that these statutes were formulated specifically for manganese nodules and are not applicable to any other resource.

12.7 Research Ships and Ship Building in India

There are several research ships currently available for oceanographic research and marine mineral exploration in India, owned by the Geological Survey of India, the National Institute of Oceanography, and the National Institute of Ocean Technology (Figure 12.2).

In addition, India now has a fleet of 16 research ships, starting from ORV *Sagar Kanya*, mainly used for the survey of nonliving resources, especially manganese nodules in the Central Indian Basin, and FORV *Sagar Sampada* was procured for the investigation of living resources (see also Chapter 2). This vessel is also ice strengthened to enable it to provide support to Antarctic expeditions. *Sagar Sampada* was constructed in Denmark. India's oldest research vessel RV *Gaveshani*, which was used to perform multidisciplinary work, was decommissioned in 2010. This ship has played a useful role in the development of India's offshore oil fields, mainly through survey and position fixing. In 1981, *Gaveshani* collected the first sample of manganese nodules from the Indian ocean. It is an indigenous ship, having been converted from a hopper barge by the Calcutta-based Garden Reach shipyard. The fourth research vessel, *Samundra Manthan*, an old fishing vessel, converted and refitted for scientific surveys of nonliving resources, is responsible for the survey of the EEZ of India.

India needs a strong shipbuilding industry for economic reasons. With the Indian peninsula having a coastline spanning 7516.5 km and covering 1197 islands, India's shipbuilding capabilities have not kept pace with its (a) economic development, (b) market demand, and (c) human resource potential. It needs to be pushed and allowed to grow at a rapid pace. The Indian shipbuilding is mainly done by 27 shipyards comprising of eight in the public sector (six yards under central government and two under state governments) and 19 private sector shipyards as shown in Table 12.6.

The present fiscal and statutory rules on shipbuilding in the country are heavily loaded in favor of export and discourage construction of ships by Indian yards for the Indian flag. It can be observed from the discussion so far that India has the potential grow in the shipbuilding sector in a healthy manner. The country needs to recognize shipbuilding industry as strategic industry. Simple taxation policies with a fully empowered regulating body for quick decision-making will aid its growth (Table 12.5).

FIGURE 12.2
NIOT's ORV *Sagar Nidhi*. (Courtesy: NIOT.)

TABLE 12.5

Predicted Growth of Shipping Industry in India

Volume	2006–2007	2007–2012	2012–2017
Order Book (Mn DWT)	1.3	5.00	18.00
Global Order Book (Mn DWT)	231.2	231.2[a]	231.2[a]
India's Share of Global Order Book	0.4%	2.2%	7.8%
Delivery (Mn DWT)	0.65	2.50	9.00
Turnover (US$ Billion)	0.65	2.50	9.00
Shipbuilding Industry % of GDP	0.04%	0.16%	0.27%
Total Employment	12,000	78,000	2,52,000

[a] The global order book is likely to decline after about 2010 onward and in that event the global share will increase.

Higher growth is needed for capacity to be increased to meet the plan target of $2.5 billion (Rs 18,000 crore) and annual turnover of 5 million DWT with a global share of 2.2%. This would result in growing 400–500 small and medium size ships in the entire period of five to six years (Tables 12.6 and 12.7).

12.8 Techno-Business Opportunity in the Indian Ocean

The oceans cover about 71% of the earth's surface and have played a crucial role in the origin of life on earth. As human beings appeared on the earth, they adapted to several

TABLE 12.6

Public Sector Ship Building in India

Name of Yard	Type of Vessel	Max. Length of Vessel which can be Built (meters)	DWT
1. Cochin Shipyard Ltd., Kochi 1972	All types up to 1,10,000 DWT	250	110,000
2. Hindustan Shipyard Ltd., Vizag, 1941	All types up to 80,000 DWT	240	80,000
3. Alcock Ashdown, Bhavnagar, 1994	Medium	90	5000
4. Shalimar Works, Kolkata, 1981	Small	55	1500
5. GRSE, Kolkata, 1960	Naval Ships	160	26,000
6. Goa Shipyard Ltd., Goa, 1967	Naval Ships	105	1200
7. Hooghly Dock and Port engineers Ltd., Kolkata, 1984	Small Ships	85	1000
8. Mazgaon Dock Ltd., Mumbai, 1934	Naval Ships	190	27,000
	Total DWT		254,700

TABLE 12.7

Private Sector Ship Building in India

Name of Yard	Type of Vessel	Length (Mtrs)	DWT
1. Elite Shipyard, Varavel, 1981	Fishing boats, wooden vessels	18	0
2. PS & Company, Vizag, 1996	Small ships, Barges	12	1000
3. Dempo, Goa, 1963	Small ships, Barges	85	3500
4. ABG, Mumbai, 1985	Small ships	150	15,000
East Coast Boat Builders, Kakinada, 1969	Not available		–
5. Bharti, Mumbai, 1976	Small ships	125	10,000
6. Chowgule & Co. Goa, 1965	Small ships, barges	100	3300
7. Alang Marina, Bhavnagar, 1987	Small ships	100	2000
8. Empreiteiros Gerais, Goa, 1962	Barges	75	1000
9. Sesa Goa, 1984	Small ships, Barges	80	3500
10. AC Roy, Kolkata, 1969	Boats, Barge, Small ships	65	1500
11. Bristol Boats, Aroor, 1973	Boats	20	100
12. Tebma, Chennai, 1956	Small ships	70	5000
13. Wadia Boat Builders, Balimora, 1991	Boats	46	0
14. Corporate Consultancy, Kolkata	Boats	40	0
15. NN shipbuilders, Mumbai, 1975	Boats, Barges	60	0
16. Western Marine Eng., Kochi, 1983	Boats, Barges	45	350
	Total DWT		26,750

habitats and also explored new horizons. Later people learned to master travel through the oceans. New continents were explored and brought under international trade and commerce. For about a millennium, mastery of travel over the oceans and command of the coastal regions were crucial for industrial and business development of the more powerful nations, which controlled the other nations as their colonies. Despite the crucial/military role played by the ocean and the coasts, their importance for has yet to be fully understood till recently. Only recently, when humanity is faced with complex problems such as population growth, environmental degradation, and fear of depletion of natural resources, attention has returned to the oceans. In addition to mineral resources and hydrocarbons from the seabed, there are other options.

On a long-term basis, the ocean is being looked at as a possible human habitat of the future, when population growth will demand more and more living space. There have been a few successful projects on building artificial islands. Because of various techno-economic considerations and possibly due to different reasons, repeated efforts to enlarge techno-business opportunities have *not* made much progress. There are, of course, several short-term and medium-term interests in the oceans and coastal regions: as a source of rich protein food; a source of new medicines; a source of energy (gas, oil, thermal and wave energy); as a dumping place of the wastes produced by human activities on the land; as a place of leisure and comfort, and so on. Since human activity has its implications for the natural balance, which has developed over decades, there are serious concerns regarding possible rapacious use of the ocean resources. Severe opposition has come from several groups of environmentalists to any intensification of the commercial exploitation of the ocean.

This section attempts to illustrate a few techno-business opportunities available, and it is hoped that industry and business leaders, scientists/technologists would work together with other public-interest groups, environmentalists, legal persons, financiers, planners, and administrators, to choose a few sustainable opportunities for the Indian Ocean. This would give strength for some futuristic areas in providing rich business and other potential strategic opportunities. This has been done with a view to stimulate interest in the large number of stakeholders referred to above, because a good understanding of the various possibilities and constraints by all of them is required to ensure sustainable growth in the ocean sector.

Ocean studies of the yesteryears, which concentrated on the study of winds, currents, sailing conditions, and navigation, have changed their scope considerably. Present-day studies not only include communication among whales and dolphins, and the use of plankton and krill as a source of protein, but also the use of minerals in the seabed, the use of desalination plants to provide freshwater, and the study of deep-diving submersibles, which could enable humans to descend to the ocean bottom to examine it. A whole range of scientific and engineering disciplines are being applied in ocean science and technology. An understanding of the environmental issues as well as a better understanding of the local, regional, and global ecosystems has added new aspects to ocean science and technology. The business opportunities therefore available through the oceans and the coastal regions would create a need for new knowledge. These techno-business opportunities are of short-term and long-term duration.

12.8.1 Short-Term Opportunity

The oil and gas resources will perhaps be at the top of the list as returns from this sector can be very high. In the international scene, countries like the UK and Norway have benefited considerably from their offshore oil industry. India Ocean has its Bombay High. Basic engineering capabilities to explore the offshore areas and to extract oil and gas resources from the seabed are available in the countries in the IOR. The procedures for the entry of foreign companies have also been simplified. Since it is likely to be an area of rapid growth in the coming years, it is also essential that the technologies to take care of environmental concerns (avoidance of marine pollution, etc.) are built into the newer activities. As the nearshore and coastal regions are explored and exploited, the trend shall be to further go down into the deeper parts of the sea, demanding more challenging engineering.

An important techno-business area is the use of the oceans for waste disposal. The wastes dumped into the oceans range from municipal wastes to chemicals and effluents from major industries. There are also, of course, oil spills. A brief survey of marine pollution in India is available in Glasby and Roonwal (1997). One important reality is that the disposal of wastes in the oceans will continue to be the easy and economical method for many years to come. In the present-day consumption styles and industrial patterns, it is difficult to conceive how the enormous quantities of wastes and effluents that come out, and which need to be moved away from human habitats, should be dealt with. Thus, there is likely to be continued pressure on oceans for quite some time to come. In recognition of this reality lies a techno-business opportunity. When scientists, technologists, and business groups join hands, it is possible to innovate, design, and operate several sewage and waste treatment plants that can greatly reduce on an economic level the waste loads going into the sea; the wastes can be recycled even while dumping the relatively cleaner residues into the sea. Questions remained whether one does such recycling at the source in the waste-producing plants themselves or after they collectively become a huge collection as they flow toward the sea. Perhaps both options and in fact a mix of these would need to be adopted. Even then, action near sea-coasts will be required because many small and uncontrolled operations may continue for many more years leading to different types of effluents streaming toward the sea.

12.8.2 Long-Term Opportunity

Another important area would be the use of the ocean's biological resources. These are not merely the protein foods coming from fish and shrimps but also the various other forms of chemicals, biochemicals, and microorganisms even from the "hot vents," which are likely to lead to newer medicines. This is a source of new genetic materials leading to engineering of better agricultural and biotechnological products. This area is going to be highly research-intensive and may become a rich source of business in the medium term. The IOR nations have long coastlines and, therefore, a large EEZ (Figure 12.1) and a good scientific and technological base could rapidly utilize these natural resources of with immense biodiversity. It cannot afford to wait for others to invent new products or processes, because it is likely that those who invent in the United States and Europe are investing heavily in such technological developments and countries like Japan and China may soon follow suit, will obtain the various forms of intellectual property right and, more important, have the advantage of an early entry into new lines of business and commercial operations. Mass production of new products based on genetically engineered material is likely to swamp the markets in an even more all-pervasive way than electronics products based on very large-scale integration (VSLI) did. This is a technological possibility for the not too distant future. World technologies, their commercialization, marketing, and protection are advancing at a very rapid pace. Decision makers and business houses need be made aware of the deeper implications of such a possibility. If at least some leverage in a few critical areas through our own core strengths is not acquired, the region shall be left lagging far behind others. The Indian Ocean provides an excellent opportunity.

Some other major opportunities such as ocean energy systems, the use of the sea and especially the coastal zone for laying telecommunication cables, and for transport, the use of ocean and coastal zone for recreational activities and tourism. The use of coastal waters for domestic transportation of goods and materials is a distinct possibility. Technologies of their transport and handling are simple. With rapid economic growth, with the possibility of the growth rate of GOP is transportation essential. Techno-business opportunities, with a value addition of environmental care, can give a leading edge to all business houses in developing and operating such services. If they do not take initiative, foreign companies may come in to provide such transport and port handling facilities. Protection and management of estuary offer scope for business opportunities to industry.

There are several possibilities arising out of these opportunities that would imply the need for detailed surveying, mapping, and providing of geographic information system (GIS) based data on the coasts, and beyond the coasts, especially in the EEZ zone. Remote sensing and various other types of sensor data and exploration data would have to be merged to provide such an information system. Weather forecast services may be superimposed on a daily basis. These activities provide immense, techno-business opportunities. Many technologies are available internationally and also in Indian Ocean region. There are also a few competent entrepreneurs in the region who can deal with mapping and surveying and present the data in electronic formats. Such information and computer simulation would also be useful for environmental protection.

Broadly, the oceans and coastal zones provide several techno-business opportunities. Such opportunities can be sustained only when environmental considerations are built into the early stages of the project. Fortunately, the current technological knowledge and scientific understanding provide an opportunity for such an early integration of efforts. The potentials for technological innovation and commercial business opportunities are immense in the Indian Ocean. The business and the research efforts have to identify and implement the required correction. Since the world is moving fast and those who are slow in action often stand to lose. Therefore, the Indian Ocean offers large opportunities, which can help grow the IOR nations for the benefit of the people. This is one way to have better economy and generate employment, wealth, and use the resource of the Indian Ocean for improving the quality of life of the people.

References

Amann, H., 1989. The Red Sea pilot project: Lesson for future ocean mining. *Mar. Mining*, 8(1): 1–22.

Amann, H., 1990. Outlook for technical advances in environmental requirement in hard mineral mining on the deepsea floor. *Mater. Soc.*, 14: 387–386.

Backer, H., 1980. Erzschlamme. In: W. Scholt (ed.). Die Fahrten des Forschungsschiffes Valdivia 1971–1978. Geowissenschaflliche Ergebnisse. *Geol. Jahrb.* 30: 77–108.

Black, J. R. W., 1980. The recovery of metals from deep sea manganese nodules and the effect of the world cobalt and manganese markets. MIT, Report.

Blissenbach, E. and Nawab, Z., 1982. Metalliferous sediments of the seabed: The Atlantis-II Deep deposits of the Red Sea. *Ocean Yearbook* 3: 77–104.

Chung, J. S., 2003. Deep ocean mining technology: Learning curve. In Proc. ISOPE—Ocean Mining Symp. pp. 1–16. International Society for Offshore and Polar Engineers, Tsukuba Japan.

Cruickshank, M. J., 2002. Mineral resources potential of continental margins. In: C. A. Burk and C. L. Drake (eds.). *The Geology of Continental Margins*. Springer, Berlin, pp. 965–1000.

Donges, J. B., 1985. *The Economics of Deep-Sea Mining*. Springer, Berlin, p. 378.

Glasby, G. P., 2002. Deep seabed mining: Past failure and future prospects. *Mar. Georesourc. Geotechnol.* 20: 161–176.

Glasby, G. P., and Roonwal, G. S., 1997. Marine pollution in India: An emerging problem. *Curr. Sci.*, 68: 495–497.

Herzig, P. M., 1999. Economic potential of sea floor marine sulfide deposits: Ancient and modern. *Phil. Trans. Royal Soc. London*, A357: 861–875.

Hsu, K. J., and Theide, J. (eds.). 1992. *Use and Misuse of the Seafloor*. Dahlem Workshop Reports 11, 17–22 March 1991. John Wiley, Chichester, pp. 440.

Indian Mineral Yearbook, 2012, Indian Bureau of Mines, Govt. of India, Nagpur.

International Seabed Authority, 2001. Law of the Sea, 484 pp.

International Seabed Authority, 2012. Basic Text, 2nd Ed. 178.

Laughton, A. S., 1996. Responsible ocean exploitation. In: S. Z. Qasim and G. S. Roonwal (eds.). *India's Exclusive Economic Zone*. Omega, pp. 18–35, 238.

Mero, J. L., 1965. *Mineral Resources of the Sea*. Elsevier, Amsterdam, 312.

MIT Report, 1978. A cost model of deep ocean mining and associated regulatory issues. J. D. Nyhard, Lantrim A., Capstoff A. D., Kohler, D. Leshow. MIT sea grant programme, MIT Cambridge, 163 pp.

Qasim, S. Z. and Roonwal, G. S., 1998. The exclusive economic zone. In: *Living Response of India's Exclusive Economic Zone*. Omega, pp. 1–4, 140 pp.

Rona, P. A., 1988. Hydrothermal mineralization at oceanic ridges. *Can. Miner.*, 26: 431–465.

Rona, P. A., 2008. The changing vision of marine mineral. *Ore Geol. Rev.*, 33: 618–616.

Rona, P. A. and Scott, S. D., 1993. Preface pp. 1933–1976. In: Special issue on seafloor hydrothermal mineralization: new perspectives, *Econ. Geol.* 88(8): 1993–2295.

Scholten, J. C., Stoffers, P., Gorbe-Schoneberg, D., and Moammar M., 2000. Hydrothermal mineralization in the red sea. In: D. S. Cronan (ed.). *Handbook of Mineral Deposits*. CRC Press, Boca Raton, FL, pp. 369–395.

Scott, S. D., 2008. Mineral deposits in the ocean: Second report of the ECOR Panel on marine mining (September 2008), 36 pp.

Glossary

Abyssal: Deep ocean environment on the ocean floor

Atol: A curved reef which is mostly submerged

Authigenic: Produced in situ

Banks: An elevation in sea or river bed

Benthos: Flora and fauna of seafloor

Benthic: On the seafloor

Biomass: Mass of the living organisms in a given volume

Benificiation: Treatment of raw material to improve the concentration of metal in it

Continental Margin: Submerged prolongation of landmass into the sea. It consists of shelf, slope, and rise

Continental Shelf: Shallow submarine plain with varying width, forming a border to the continent, and ending in a steep slope (the continental slope) into the oceanic plain (abyssal plain)

Coral: Composed of hard part of skeletons of a variety of marine plants and animals

EEZ: Exclusive economic zone; the area extending from the shoreline up to a distance of 360 km (200 nautical miles) on which the coastal state has sovereignty for the exploration and exploitation of all resources

Ferroalloy: Iron-rich refined product, containing other valuable metals

Hydrothermal: Formed through the actions of heat and water

Ferromanganese: A ferro alloy containing iron, manganese, and carbons

Mine Site: Area needed to sustain polymetallic nodule mining operation

Mineable Proportion: Areas where mining will be adopted, excluding the areas where there are no nodules, or where recovery of nodules is not possible due to uneven terrain

Nodule Abundance: Average weight of the nodules per square kilometers

Ore Grade: Containing metals in high enough concentration to be a potential ore

Pelagic: Of or performed in the open ocean

Phosphorite: Aggregation of main calcium phosphate minerals, formed on a seabed or on land

Plate tectonics: Concept explaining geological features through the action of lithosphere plates that either move or collide to form new crest or denudation of existing crest. It may lead to mountain building processes

Reserve: The part of the resource, which is economically exploitable with current technology

Resource: Mineral deposit that are known or inferred to exist

Seamounts: A mountain/dome rising abruptly from the seafloor. A submerged seamount with a flat top is known as a Guyot

Siliceous: Ooze soft deposit containing silica

Sulphide: A mineral or ore deposit in which the metals occur in combination with sulfur

Turbidity Current: Large moving mass of water containing suspended sand silt and clays

Water Mass: A large mass of water identifiable through its temperature and salinity

Index

Printed and bound by CPI Group (UK) Ltd, Croydon, CR0 4YY

24/10/2024

01778286-0014